Bob Levine
3 June 80

Vineyards in England and Wales

also by George Ordish

WINE GROWING IN ENGLAND
(*Hart-Davis*)
THE GREAT WINE BLIGHT
(*Dent*)

Vineyards in England and Wales

GEORGE ORDISH

FABER AND FABER LIMITED
London · Boston

First published in 1977
by Faber and Faber Limited
3 Queen Square London WC1N 3AU
Reprinted 1978
Printed in Great Britain by
Redwood Burn Ltd, Trowbridge and Esher

British Library Cataloguing in Publication Data

Ordish, George
Vineyards in England and Wales.
1. Viticulture—England
I. Title
634'.8'0942 SB397.G7

ISBN 0–571–10928–4

To the memory of Edward Hyams

Contents

Preface

I am grateful to a large number of people for much help with this book. First of all to Julian Jeffs, the general editor of this series: then to the authors, living and dead, mentioned in the Bibliography, and thirdly to many other people and institutions too numerous to mention, but I particularly thank the following: Messrs. F. A. Arnold, J. G. and I. M. Barrett (Secretary of the English Vineyards Association), R. Barrington Brock, James Bleasdale for checking the daylight calculations and arranging a tour of vineyards, S. J. A. Evans, the late Edward Hyams (a great expert), Anton Massel, W. B. N. Poulter, my nephew Jeremy Putley, Joan and Michael Saunders for news from Lincoln and for other items, Jack Ward (Chairman of the English Vineyards Association) and the Librarians (and staffs) of the Bodleian Library (Oxford), British Museum (Natural History and Bloomsbury), Rothamsted Research Station, St. Albans Public Library, Tropical Products Institute (London), and Verulamium Museum (St. Albans). I also thank the many British vignerons I have visited for much help. Finally I thank my wife for checking the manuscript and much assistance in preparing this book.

The measurements in this book are mostly in British terms, but as the country is gradually going metric (a system anathematized by George III as being based 'on those damned dots'), the following conversion factors are mentioned: 1 acre = 0·404 hectare, 1 imperial gallon = 4·54 litres, 1 lb. = 454 grammes, degrees Farenheit — 32 × $\frac{5}{9}$ = degrees Centigrade. The analytical data are already in metric figures.

G.O.
St. Albans.
January 1976

CHESHIRE
CLWYD
GWYNEDD
STAFFS
SALOP
WEST MIDLAND
POWYS
Llanarth
Shelsley Beauchamp
HEREFORD
Suckley
Clyro
Bodenham
& WORCESTER
DYFED
Newent
GLOUCESTER
Lamphey
Glascoed
GLAMORGAN
GWENT
Thornbury
Ashton Keynes
AVON
Bradford on Avon
Dunkerton
Melksham
Axbridge
Wedmore
WILTS
North Wooton
SOMERSET
Pilton
Dulverton
Shepton Montagu
Washfield
Stapleton
Broadclyst
Wimborne
DEVON
Stoke Abbot
DORSET
CORNWALL
Stoke in Teignhead
Roseworthy
St. Keverne

Stragglethorpe
DERBY
NOTTS
LINCS
LEICESTER
North Elmham
NORFOLK
Hempnall
East Harling
South Kilworth
Brampton
CAMBS
Fressingfield
WARWICK
NORTHANTS
Wilberton
Barningham
Hoxne
Halesworth
Thurston
Bury St. Edmunds
SUFFOLK
Kelsale
Wickhambrook
Gamlingay
Wetheringsett
Otley
Linton
BEDS
Cavendish
Duddenhoe End
Langham
BUCKS
Ardleigh
Felsted
Colchester
Shipton-under-Wychwood
Weeley
HERTS
Kingston Bagpuize
ESSEX
Berkhamsted
OXFORD
Purleigh
Gt. Burstead
BERKS
Pangbourne
Chieveley
Purley
LONDON
Bradfield
Kintbury
Finchampstead
Lullingstone
Inkpen
Pewsey
Faversham
Nettlestead
Hook Heath
Matfield
West Peckham
KENT
SURREY
HANTS
Ulcombe
Wye
Godalming
Lamberhurst
Biddenden
Bishops Sutton
Goudhurst
Haywards Heath
Ticehurst
West Tytherley
Bolney
Uckfield
Heathfield
Hambledon
Plumpton Green
Broad Oak
Singleton
Newick
SUSSEX
Beaulieu
Rowlands Castle
Westfield
Blackfield
Arundel
Rodmell
Hailsham
Hayling Island
Roedean
Cranmore
Sandown
Vineyards of England and Wales
5 acres and over
Under 5 acres
0 10 20 30 Miles
NSH

I

Introduction

It is not as difficult as might be imagined to grow grapes in England and Wales and then to make good wine from them. In the first place our much maligned climate is nothing like as bad as it is often said to be and compares favourably with that of some of the north European vineyard areas. Looking at the statistics one sees that the figures from weather stations near our vineyards, from the wine-making point of view, are in some respects sometimes better or about the same as those obtained from the northern European wine areas.

The important growing and ripening months are June, July, August and September. Statistics show that on average (thirty years) for these months Sandown, Isle of Wight, has considerably more sunshine and much less rain than Rheims in the Champagne, France, or Frankfurt-am-Main in the German wine area. Bournemouth and Clacton-on-Sea also have much less rain than Rheims and Frankfurt-am-Main and about the same amount of sunshine (see Appendix I, p. 157). Mr. W. B. N. Poulter, who has a vineyard in the Isle of Wight, has also studied this subject and points out that the actual climate (the 'micro-climate') can vary considerably from that recorded by a nearby meteorological station. This subject is discussed more fully in Chapter 4.

Nor is Britain exactly in the Arctic Circle, as some of our detractors seem to imply. Much of Germany is north of London and at least one English vineyard, at St. Keverne, Cornwall, is actually south of the Rheinburgengbau plantations at Ahrweiler on the banks of the River Ahr (Germany); St. Keverne is 50° 2′ North latitude while Ahrweiler is 50° 32′ North. Nevertheless, the northern French and German sites are on the whole more southerly than the English ones, a fact making site selection yet more important for success in Britain.

As vines have been grown in Britain and wine made from the resulting grapes almost continuously since at least the time of the Romans, it is of interest to enquire why any surprise should still exist as to the possibilities of English vineyards and the good wines they produce. Considering the last seventy-five years there seem to me to be five main reasons for this belief. First, there is the general denigration of our climate, already mentioned; all too many of us tend to regard the summer as 'good' or 'bad' according to the weather we happened to experience during our annual holiday. Secondly, there are the trade interests: until fairly recently the brewers discouraged wine-drinking. It was put about that the mystique surrounding wine was complex and that acquisition of that knowledge, let alone the wine itself, could only be undertaken and appreciated by the long purses and refined palates of the rich. An offshoot of this was that at the turn of the century a wine merchant was considered a gentleman, whereas a grocer was not—a curious anomaly.* For instance, about that time Sir Thomas Lipton was blackballed on a vote for his admission to the Royal Yacht Club, ostensibly on the grounds that he could not sail a yacht, until his friend the Prince of Wales (later Edward VII) threatened to resign from the Club. The founder of the big grocery chain was then accepted. Thirdly, there was the government interest. Vineyards until recent times tended to be small, particularly in England. British governments have always raised revenue by taxing alcoholic drinks (they now tax soft drinks as well) and it was far easier to do that by an excise duty on a relatively small number of breweries, distilleries and on consignments at the ports, than to send tax collectors round hundreds of small wine growers. Fourthly, from the commercial point of view, cultivating a vineyard, while full of interest, is labour intensive, difficult and extremely hard work. Apples or plums would give as good a cash return as wine, with much less effort. Fifthly, the two American vine mildew diseases, reaching Europe in the nineteenth century, added another setback. Neither was difficult to control, but to do so meant additional work and an extra expense, and the sight of old-established vines on many a mansion

* cf. *Man and Superman*, Bernard Shaw (1901). John Tanner has been held up by brigands in Spain and is brought to their camp:

MENDOZA: Allow me to introduce myself: Mendoza . . . I am a brigand. I live by robbing the rich.

TANNER (*unperturbed*): I am a gentleman: I live by robbing the poor.

wall slowly dying from the powdery mildew discouraged potential vignerons from planting up.

Today things have changed considerably. Wine is now drunk by pretty well all the drinking classes; brewers vie with one another in offering a choice of wines in their pubs. Amateur wine makers, using all sorts of fruits and vegetables, abound. There is an English Vineyards Association in existence with nearly 400 members and there are about 600 acres of open-air vineyard vines now growing in the country.

There are at least two northerly vineyards in Britain. One is the, now revived, old vineyard of the Bishop's Palace at Lincoln, which picked its first (modern) crop in 1975; the other is at Stragglethorpe Hall, between Sleaford and Grantham in the same general area. The Old Palace vineyard is at 53° 14′ North latitude, a considerable distance north from the alleged northward limit to vine cultivation—51° N. This first crop shows every promise of producing a fine wine.[58*] The vineyard obviously is on a good site. It is also an example of the friendliness of vignerons. The plants for the Old Palace vineyard were given free by the Wine Club of Neustadt, Germany, with which city Lincoln is 'twinned'. Neustadt, the Upper Palatinate vineyard area producing large quantities of *café-wines*, far from wanting to reduce competition, is encouraging the Lincoln effort and sent some of their wine to celebrate the first Lincoln harvest.

Wine is of great help in maintaining health and it is interesting to note that one of our leading health farms (the original one, in fact—Champneys, Tring, Hertfordshire, started in 1925) now makes wine available in most of the diet régimes for its guests. Wine is a living thing and contains certain useful vitamins, polyphenylenzymes and other constituents the body needs. It is very beneficial when taken with meals, for not only does it promote a feeling of well-being but, by slowing down the speed of feeding (taking a sip of good wine every now and then and appreciating it does this), it also aids digestion. St. Paul wrote to Timothy: 'Stop drinking nothing but water; take a little wine for your digestion, for your frequent ailments.' Needless to say, to retain health, wine consumption must not be overdone. Champneys has now an English vintage on its wine list which has been found to be popular.

* Entries in the Bibliography are referred to in this way throughout.

The dedication of the vigneron to his task is remarkable. It has been noted by historians from the earliest times and still persists, because the vine is a hard master and requires much effort from its devotees. The work is almost constant and the returns are always at the mercy of the weather. That such a labour-intensive occupation should be growing so fast in Britain provides an interesting contrast to the economies of today's world, for many labour-intensive industries are being run down—the railways, plum- and strawberry-growing for instance. Ten acres of vines (not a vast vineyard) are a full-time job for a man, and he will need additional help in tying up, picking and making the wine: we may well ask why people take up this hard work.

To my mind the answer lies above all in the fascination of the task itself. Moderately to partake of wine is a civilized thing to do; it maintains friendship. To drink your own wine—the material result of hard work, skill and intelligence—is highly satisfactory. It is work that makes the wine, and it would not be surprising were all wine growers, or at any rate the new ones, Marxists—believing in the labour theory of value—except that they are probably too busy in the vineyards to spare much time for political theory. There are two great moments that reward the new vigneron for the three years of waiting for his first vintage. The first is when he has picked his grapes, fed them into the press and set the must to ferment in his outhouse or winery. As the evening calm falls he goes into it to see if all is well. A delicious smell fills the air and he hears that magic sound—the tiny crinkle of the bubbles bursting in the vat. The noise is only just perceptible, but thrilling nevertheless—he has become one of the long line of wine makers. Wine is one of the civilizing influences of life; it helped primitive man to confront the terrors of his world—the black woods, the frightening animals and the threatening and mysterious activities of the dark gods.

The second moment, of course, is when he actually drinks his first wine and he must be careful, if it is even reasonably palatable, not to endow it with qualities it does not really possess, such will be his pride in it.

In England today two kinds of people seem to indulge in wine growing, retired men and women and comparatively young couples. Better general health and early retirement mean that many of the people who have just retired are active and by no means anxious to stop working. Not afraid of hard work in their own interests, and appreciating the qualities of wine, the retired can find great satisfaction in developing a vineyard. They are cheap labour and find more pleasure among the vines, in spite of all the cares the plants may bring, than from working in any other capacity available to the recently retired.

The young who take to vine growing seem to be those who dislike a steady nine-till-five job and welcome independence. Many of these are working part-time while developing their vineyards and are extremely enthusiastic about them, experimenting and improving their methods all the while, looking forward to the day when the vineyard will be their full-time job and livelihood. Young people have two considerable advantages over the retired: provided they do not smoke, they are likely to have a better palate (though perhaps experience does count for something) and thus are better placed to judge wine; secondly, being young, they are likely to have more time to do it in.

There is also a remarkable spirit of good-fellowship among wine growers; they all help one another both with practical aid, such as picking the grapes, and with long discussions on methods and results. This social solidarity is also important.

The growing of our own wine in England is obviously economically advantageous to the country, as it helps to correct our balance of payments problem. In 1974 the United Kingdom imported 44 million gallons of light wine, worth about £83 millions (c.i.f.—that is, before duty was paid). Unfortunately for the wine grower our present government do not seem at all anxious to foster this new crop; they tax home-produced and imported wines at the same rate of duty (but see also p. 142).

Another encouraging factor in favour of English wines is that many fruits have a much better flavour at the edges of their respective cultivation zones. Spanish oranges are much better to eat than North African ones; English apples are better than French; French olives are superior to Italian and so on. A slow growing season seems to develop flavour in fruit to a remarkable degree, and this may well be a reason why some carefully made English wines can rival their continental counterparts in quality.

Wine is a civilized thing to have and enjoy—a promoter of conversation and good talk. Listen to an extract from a popular eighteenth-century work; it had gone to eight editions by 1757:

'*The Countess.* I am not so much surprised at the Production and agreable Flavour of Wine as I am at its beneficial effects. Other Liquors, whether natural or artificial, as Beer, Cyder, Tea, Chocolate and Coffee create Silence and Serious Airs, for the generality, and consign those who drink them, to a melancholy Cast of Mind. . . . But it is the peculiar Privelege of Wine to introduce Variety and Joy, wherever it appears.'

Spectacle de la Nature, translated by S. Humphreys.[103]

2

The history of English and Welsh vineyards

Writers on English vineyards in our present mid-century revival, from the pioneering book of Edward Hyams (1949) to the present day, have accumulated a considerable body of evidence on the early English vineyards.[48] A book I myself wrote in 1952 brought together a certain amount of historical information.[78] Many of the later writers seem to copy from the earlier works, the admirable dictum 'check your references' being somewhat forgotten. I have been fortunate in coming across an interesting article on the early vineyards by Samuel Pegge (1704–96), which appeared in Volume I of *Archeologia*, a journal of great appeal at that time, for two more editions of the first volume came out quite quickly, the last (the third) appearing in 1804.[85] It provides a few more clues, in addition to those in Edward Hyams's book, and these and some of my own discoveries are dealt with below. I have been able to verify most of Pegge's references.

Samuel Pegge was the son of a wool merchant and became a pensioner and scholar of St. John's, Oxford. He entered the church, held a succession of parishes as vicar or rector and wrote extensively on antiquarian themes—Anglo-Saxon jewellery, King Alfred, King John's death and English vineyards among many other subjects.[24] He was a great scholar, historian and antiquarian and we cannot do better than follow his vineyard comments, though with some interpolations from the other sources mentioned, in order to preserve a rough historical sequence, together with some discussion of various points raised. This means that a certain amount of to-ing and fro-ing must be used instead of a strict chronological order.

To begin with, Pegge points out that Julius Caesar does not

mention vines in Britain and that Tacitus specifically denies us that plant. He then comes to Domitian. It is now fairly well known among wine lovers that the Emperor Domitian (reigned A.D. 81–96) prohibited the planting of vines outside Italy; Pegge points out that the Emperor also prohibited the establishment of new vineyards within Italy, and decreed that existing vineyards outside the homeland were to be reduced to half their size and no new plantations were to be made. His reasons for this measure were political and economic. In the colonies, he maintained, particularly in Britain, drunkenness led to sedition and revolts. In passing we may note that Pegge quotes Prideaux, saying: 'This seems to be the ground of that piece of policy in Mohammed, who denied his disciples the use of wine for this reason amongst others.' Domitian also felt that in Italy wine was so profitable that the growing of corn was neglected and he feared a food shortage. 'Bread and circuses', after all, was the guiding principle for keeping the Roman population quiet.[54, 109] The Domitian edict naturally also helped the existing producers, maintaining prices at a high level, like the Common Market policy of today.

The fact that the edict outlived Domitian by about two hundred years shows that policy to have been acceptable on the whole. As for the Emperor, in spite of his elaborate precautions—in his palace polished marble walls acted as mirrors so that he could always see behind him—he was assassinated in A.D. 96, aged 45. The edict was naturally disobeyed in many parts, though not in Britain, and in Asia it was cancelled. In view of the fact that the vine (*Vitis vinifera*) is indigenous to the Near East and that wild vines abounded there, the edict would have been almost impossible to enforce.

In all probability the edict had the opposite political effect to that intended. The 'barbarians' outside the Empire much appreciated wine and one of their main reasons for attacking Italy was to get at the wine, rather than at the beautiful women, as we have been led to believe. Moreover, it was just as easy to get drunk on mead and beer. In fact it always seems to me that the beer-drunk individual is aggressive whereas the over-indulged wine drinker is mild and maudlin. Pegge quotes the historian Ammianus Marcellinus (died about A.D. 390), who pointed out that the Gauls and Britons had recourse to other equally intoxicating liquors: 'As a nation they were very desirous of wine and made numerous drinks similar to it.'[85, 60] Had the barbarians been encouraged to grow their own vintage the Roman Empire might not have been under such hostile pressure for so long.

In A.D. 276 the Emperor Probus repealed Domitian's edict. Several reasons for this action suggest themselves: the measure was almost impossible to enforce, wine growers were too prosperous and arrogant, prices were far too high and should the edict be repealed there would be a demand for Roman technical experts in the 'underdeveloped countries'. The presence of these experts, sought for and welcomed by the barbarians, would enable the Emperor to secure accurate news of what was happening in those far-away places, as is still the case with the export of technicians today. Probus, moreover, was the son of a peasant and a generous and likeable man. He may well have recalled his youth and have felt that good wine should not be the exclusive perquisite of patricians but available to all. The planting and working of vineyards also helped in the unemployment problem, particularly that of ex-soldiers. The measure was opposed by Proculus and Bonosus, two army officers who both aspired to the Imperial Crown, and both seem to have enlisted the support of the existing wine growers, who had a considerable vested interest in maintaining a restricted acreage. Proculus and Bonosus were unsuccessful and were hanged. It is somewhat ironic that it was the planting of vines that led to Probus's death. He was killed (A.D. 282) in a mutiny of his own men, who rebelled because they had been ordered to do public work, including the planting of vineyards, rather than led to get rich by means of battle and loot.

The upper-class Romans in Britain drank wine: there is hardly a villa excavated today that does not yield a wine amphora. An example is that given by Pegge, who reports the find of a wine vessel by a Dr. Musgrave at Devizes.[85, 72] The Verulamium museum at St. Albans has many such amphorae. At first all these wines were imported and thus very expensive, but after Probus's edict vineyards were planted around many an estate. For instance, Mr. I. A. Richmond reports the finding of vine stocks at a villa excavated at Boxmoor, Hertfordshire, and Verulamium may have had its vineyard too (see page 30).[92] The Roman soldier, like the British sailor his rum, had a ration of wine as his right. It was called *posca* and was imported at first in amphorae and later in casks. Dr. Smith [101] maintained that *posca* was vinegar, but it seems to me more likely to have been wine, frequently souring and on its way to becoming vinegar. Whether wine or vinegar, the Roman commissariat must have had every incentive to produce this drink locally and do away with expensive imports. When the Romans withdrew from Britain (around 400–436) the vineyards were neglected, according to Charles Singer, but did not disappear entirely.[99] Medieval viticulture was but a slow rediscovery of the ancient technique, and by the ninth century it was beginning to flourish again.

It must be remembered that in classical times, particularly in Greece, the strong local wines were usually diluted with water, and the amount of water to be used, often three parts of water to one of wine, was a frequent subject of discussion. The kind of water used was important too. In summer a host would send a man with a donkey fitted with panniers into the mountains. The man looked for snow pockets on the northern slopes, filled his baskets and returned in time for the feast. The water was thus about as expensive as the wine and dilution was not just a way of making the wine go further.[96] Sea water was used in some cases, which must have made a strange mixture. In the colonies it was obviously advantageous to grow your own wine. This colonial imitation would not be as heavy or as alcoholic as the resinated wines of Paros and Santorini and would not need dilution. Indeed even in Rome, if Trimalchio's feast is anything to go by, the later Romans did not tend to put water in the wine.

Any Roman plantations in the more northerly colonies—northern France, the Low Countries, Rhineland Germany and Britain—would all have been made with plants from Italy and many of them, if not most of them, would have been of varieties unsuitable for the new climate.

Viticulture somehow inspires devotion in its practicians everywhere and gradually the northern vignerons would have noticed that certain new plants, from seedlings and bud sports, had advantages over the original Italian stocks—earlier ripening was an obvious one. The growers would then have replanted their vineyards with the new kinds. In some of the old vineyards the plants may have been grafted over to the new cultivars—an art that was to become of vital importance a thousand years later (see p. 68). Growing grapes and making wine are skilled tasks and the services of experts from Italy must have been much in demand in the early days. But even these specialists had no knowledge of the principles involved, in fact nobody had until the time of Pasteur (1822–95) about a hundred years ago, and even then most of the experts were reluctant to believe the great chemist.[83]

It was the church that fostered wine making. The clergy put forward the view that vines had to be planted to secure wine for the mass and that it was easier to grow wine than to import it. Moreover, it was establishing a new industry and bettering the populace. It has been suggested that the wine was needed more for the refectory than for the chalice, because in the Roman Catholic church only the priest takes the wine, so that not much was really needed. While there is no doubt that monk, abbot, bishop, cardinal and pope, leading their austere lives, consoled themselves with wine, there was also a considerable demand for communion wine from the ninth to the thirteenth centuries, for up to this latter date both the congregation and the officiant took the wine. It was the Synod of Lambeth, 1281, that restricted the wine to the celebrant only. It may be noted in passing that one of the reasons for Henry VIII's popularity was that he 'restored the cup to the people'. In the Anglican church communion is still 'under two kinds'.[81]

The importance of wine and other alcoholic drinks (beer and mead) in monastic times may be seen from the fact that the cellarer was a most important official. So important, in fact, that he became second in command to the abbot and the administrator of the church's lands. Ada Levett has compiled a list of cellarers at St. Alban's Abbey (seventy-one names) from Adam (1151) to Robert Blakeney (1539), the case of Adam being remarkable.[57] He was illiterate, yet ran the estates so well, and presumably the cellar, that he retained the post for a long time. The practical and intelligent illiterate does have one great advantage over his literate cousin—he has not ruined his memory by crowding it with things he has read. His head is stocked with useful facts he has heard or seen. Adam the Cellarer must have been just such a person and knew who was farming well and had paid his tithes, and who had not.

In those days alcoholic liquors were important in preserving health: to drink water was dangerous, particularly in towns, because wells could so easily become polluted with sewage or dead animals. In any old cookery book the cook is always told to use 'fair water', quite a problem then. Even if the water looked and tasted 'fair' it could still have been polluted. Beer had been boiled at one point and wine had enough alcohol in it to kill typhoid germs, so both were much safer drinks.

Ada Levett also quotes from some little-known manorial deeds showing the presence of vineyards. Between 1240 and 1249 Margaret of the Vineyard at Kingsbury Manor (near St. Albans) gave 12*d*. on being recognized as the true heir of that land, as declared by John, headman of the tithing, and the said Margaret gave the lord 2*s*. inheritance fine for half an acre of land in the vineyard.[57]

In the early days it was difficult to keep wine. That simple artefact, the cork, had not been invented. Wine was drawn as needed from barrels, and as the wine flowed out from the tap, air flowed in through the loosened bung or spile hole, carrying the acetic acid bacteria with it. With plenty of air available the bacteria were soon at work, building up acidity in the wine and eventually turning it to vinegar. A cask of wine once started had to be finished fairly soon. This led to two things, an increase in consumption—'Let's have another one; got to finish the barrel'—and a demand for small barrels. The great advantage of a goatskin as a wine container was that it collapsed as the wine was drawn. It might impart a certain leathery taste to the vintage, but no air, and thus no vinegar bacteria, entered as the wine was drawn. The principle is in use to this day for draught wines in our bars. That stout, highly decorated and splendid oak barrel from which a glass of sherry or table wine is drawn for you usually has inside it a sealed plastic bag, which, like the goatskin, collapses as the wine runs out. It thus keeps the wine in its original condition and out of contact with the air. But, like the goatskin, the plastic, unless carefully chosen, can add its different, but distinctive, flavour to the wine!

Another difficulty in keeping wine in barrels is that the wine slowly evaporates through the wood (it tends to get weaker with age, not stronger as is sometimes thought; the alcohol evaporates more quickly than the water), increasing the ullage space (air) in the barrel, which gives an opportunity to bacteria and fungi to work on the wine and spoil it unless the ullage is constantly topped up with sound wine. This evaporation is the means by which wine improves in barrel, but in medieval times it is doubtful if many cellar-masters realized it. Some undoubtedly did know from experience that air somehow was the enemy of wine, but that knowledge was not at all general. The result was the complete opposite of the position today. New wine was considered to be better than old wine and the new wine made higher prices (see page 34).

Another difficulty facing the early monkish vineyards was that in the northern countries the white wines made there were, and still are, considerably better than the red, but the wine for the mass had to be red. The early wine makers had no compunction in adding all sorts of substances to their vintages and a splendid fruit for this purpose was to hand—the elderberry.

The first complete printed book entirely devoted to wine was an anonymous publication *De vino et ejus proprietate*, 1480.[23] The author is thought to have been Philippe Beroalde (or Beroaldo) (1453–1505). He was a professor at Bologna, secretary of the Republic and Bolognese papal delegate to Pope Alexander VI, the famous Borgia.

His work was, of course, considerably later than the medieval revival of vine growing in England, but it is quoted here to show that as late as the end of the fifteenth century all sorts of additions to wine were advocated and presumably used. He recommended such substances as white clays, egg-whites, vine roots, juniper roots and berries, wheat, hops, ivy leaves, pine needles, sand, willow bark, hyssop, comfrey, dittany, parsley, leek seed, nettles, powdered deer horn, and powdered rue seed. This last probably had some drug principle in it; for the author said wine treated with it made men drunk very easily. The problem of getting red wine for the mass did not exist in Bologna, of course, but the book does show that additions to wine were not then frowned upon. Some of the substances, such as white clay, sand and egg-white, were obviously clarifying agents and are still in use, while most of the other materials had some preservative action—hops for instance. What a hopped wine tasted like is difficult to imagine! In practice, though, those additives may have been better than some of today's measures with the cheaper wines, such as stuffing them full of sulphur dioxide gas in order to ensure that they remain bright, unalterable, sterile (and dead) for years, so that they should have an indefinite 'shelf life'. And this is to say nothing about other of today's additions, for example potassium sorbate (for sweetening wine).

Let us now return to the Saxons and Anglo-Saxons. That the Saxons knew and liked wine may be seen by lines in some of their old Norse sagas. *Beowulf* is an instance, a story written in the eighth century A.D. but recording events of the sixth. In *Beowulf* the Danes, after avenging a treacherous attack on them, held a lavish feast in their great hall. A poet told a story of how King Finn was killed, then, 'As mirth renewed, and laughter rang out, cup-bearers poured wine from wonderfully made flagons'.[9] From whence did this wine come? Very probably down the Rhine from Germany, though some of it could have been grown locally. That taste for wine they undoubtedly carried with them to England, where they found the natives too liked it and used it. Another reference of that period is found in Henry Sweet's studies (1885). Among them is an early ninth-century Kentish charter of one Oswulf (not presumably the King of Northumbria, the date is too late) in the Canterbury archives.[106, 18] The charter sets out how much honey and wine (among other things) is to be delivered. The *Blickling Homilies* refer to a tenth-century aesthete who 'ne drinch he win ne ealn', the latter article being ale.[11]

The Celts in England, as noted above, were consuming a certain amount of wine at the time of the Anglo-Saxon invasions, some of it home-grown and some imported, the latter paid for by the export of tin, copper and slaves. The existence of these vineyards is supported by a fairly well-known reference in the Venerable Bede's (*c.* 637–735) *Ecclesiastical History of the English Nation*, a work also noted by Pegge. Bede in his introductory chapter gave a short description of England and Ireland and their inhabitants. He wrote: 'Britain excels for grain and trees, and is well adapted for feeding cattle and beasts of burden. It also produces vines in some places, and has plenty of land and water-fowls of various sorts; it is remarkable also for rivers and plentiful springs.' He reported the presence of vines in Ireland too. 'The island abounds in milk and honey, nor is there any want of vines, fish or fowl; and it is remarkable for deer and goats.'[8]

As regards England, Edward Hyams pointed out that, though the reference to vines is very brief, the above paragraph is about the sum total of information on agriculture, and consequently the fact that vines are specifically mentioned indicates that they were of importance. Vines, of course, would be of particular interest to the saint because, anxious to spread Christianity, he would need wine for the mass. That there were vineyards in Ireland has been disputed; in fact André Simon thought that neither were there any in England and that the word *vinea* referred to orchards. This is not very likely; there are perfectly good Latin words for an orchard—*pomarium*, *arbustum*—and Edward Hyams found a plan of the Canterbury Cathedral Chapter House lands showing both a *vinea* (vineyard) and a *pomarium* (apple orchard) side by side.[48, 50] It could be argued, of course, that if Bede was wrong about the *vinea* in Ireland he might be equally wrong about them in England, but there is a considerable body of other evidence as to their existence here. Perhaps they were also in Ireland in spite of the climate. One of Ireland's great disadvantages as a wine country today did not exist in those days—the two American mildews. Those two diseases did not reach Europe until the nineteenth century; they would now flourish so much in the mild, damp Irish weather as to make vine growing there very difficult and expensive.

According to the eighteenth-century writer J. Strutt, the Saxons called October the *Wyn Moneth* (or *Monat*), because that was the time of the wine harvest.[105] If this is the case the crop must have been of considerable importance in those days for one would hardly call a month after a minor crop.

Other evidence for these ninth-century English vineyards is provided by one of King Alfred's (849–900) laws, which laid down that anyone damaging another's field or vineyard should pay compensation to the owner at an exact valuation, and by a document (955) of King Edwy (or Eadwin or Edwig) giving a vineyard to the monks of Glastonbury Abbey.[1, 28, 57, 85, 118] A photograph of this document was made into Plate 70 of Edward Hyams's book *Dionysius*. A difficulty here, though, is that according to another source, Edwy 'the Fair' (he was very handsome) is said to have been reproved by Dunstan, Abbot of Glastonbury (later St. Dunstan) either for marrying or 'keeping as a concubine' his near relative Algiva.[22] The Abbot was apparently banished from the realm by the angry King, the monastery was rifled and secular priests were put in charge of it, so one would hardly expect the King to be making handsome gifts to it. Of course, it could have been repentance or, since Edwy reigned from 955 to 959 and the transfer is dated 955, the gift was probably made before the fatal reproof was given. One hopes the vineyard was not laid waste. It is perhaps of interest to note the rest of the story. The incensed Abbot ganged up with the discontended Mercians and Northumbrians, which so worried Edwy that in 959 he 'died of a broken heart', aged about twenty. One needed to be careful of offending abbots in those days.

The vines flourished around Winchester, and Pegge quotes Somner[102] to vouch for this. He also mentions an Elizabethan historian, John Twyne (died 1581), who put forward the view that the very name of the city was derived from the vineyards there.[110] Twyne maintained that the Romans called it *Vintonia*, which became *Wintonia* in due course, and then the Saxons named it Winchester. 'A fortified town full of wines and vineyards, where the best wine in Britain grows.' Pegge points out that other authors did not agree with that derivation of the name, particularly a Mr. Baxter and Dr. Musgrave who said the etymology of *Vintonia* was obscure. Twyne goes on to point out that the Abbot of St. Augustine's at Canterbury, Kent, said that 'At one time no reasonably sized house was without its vineyard, for example Northames, Fishpole, Littlebourne, Conningbrooke, in the parish of Sellinge and a few others.' Somner thought that the street in Canterbury called Wincheape took its name from being the old wine market.

Pegge quoted William Lambarde (1536–1601) as follows: ' "History hath mention, that there was about that tyme (the Norman invasion) great store of Vines at Santlac (near to Battel in Sussex)."[56] He adds as to Berkshire: "The like whereof I have redd to have been at Wyndsore, in so moche as tythe of them hathe bene theare yelded in great plenty, which gyveth me to think that wyne hath been made long sence within the Realme; although in our memorie it be accompted a great deintye to heare of." ' Lambarde's orthography reminds me of *Artemus Ward in London*; on 'Chawcer' he remarked, 'He's the wuss speller I know of.'

With the Norman conquest and the Domesday Book (1086) we get more accurate information. There are thirty-eight vineyards mentioned in that work, well distributed over the south of the country—Berkshire, Wiltshire, Westminster, Holborn, Hatton Garden, St. Giles, Hertfordshire, Worcestershire, Essex, Gloucestershire, Kent, etc. It has been suggested that Vine Street, Piccadilly, where the famous police station is to be found, used to be the Holborn vineyard of the Domesday Book. This seems unlikely in view of the distance between the two points. It is more probable that the eleventh-century Holborn vineyard was situated in the (later) Bishop of Ely's palace garden, around what is now Ely Place, a cul-de-sac just north of Holborn Circus, and possibly the vines ran up to what is now Farringdon Road station. Vine, Castle and Peter streets were swallowed up there in the 1856 clearances and improvements.

The Ely Place area still possesses a pub called The Mitre, no doubt commemorating the old palace. The Bishop's garden ran down the gentle slope to the Fleet river, ideal for a vineyard. Queen Elizabeth I made the property over to Sir Christopher Hatton, an event commemorated to this day by the name of the jewellery trade centre—Hatton Garden. Existing and former street names in the area, such as Saffron Hill and Pear Tree Court, are evidence of the horticultural past of the district. Obviously it would have been a centre of market-gardening; it had a good soil, was watered by the Fleet River and had an enormous market for its produce to hand—the City of London. As regards the vineyard, it is perhaps even more significant that some of the south-westward facing houses in the Farringdon Road still have vines growing on them. But these charming early Victorian buildings have a doomed look about them; no doubt they and their vines will soon be gone.

Vine Street, Piccadilly, probably commemorates a vine or vineyard in the gardens of Burlington House, there in the early eighteenth century.

References to most of the Domesday Book vineyards can be found in Sir Henry Ellis's *General Introduction to the Domesday Book* (1833) and in the *Victoria County Histories*.[29, 82] Taking Hertfordshire as an example (the volume was published in 1923) references to two Domesday vineyards are found:

i. The Manor of Standon had a vineyard half a mile east of Ermine Street: it was two arpents in extent, an arpent being 0·8 acre, or $\frac{1}{3}$ hectare. The existence of this vineyard is thought to indicate the residence of a lord.

ii. At Ware a manor held by Hugh de Grentmesnil had land for 38 ploughs, meadow for 20 ploughteams, woods for 400 swine, two mills worth 24*s*., and 375 eels, an enclosure for beasts of the chase and 4 arpents of vineyards, newly planted, which shows that wine making was sufficiently successful for new vineyards to be set.

It is said by E. S. Rohde[94] that vines were so numerous in Ely that the Normans called it the *Isle des Vignes*, but unfortunately that author does not give her source and the present expert on Ely, Dean S. J. A. Evans, knows of only two former vineyards, both in the city itself—the Bishop's vineyard (still known as The Vineyard) and the Monks' vineyard opening on to Broad Lane. The Cathedral accounts show considerable imports of wine, so local production cannot have been sufficient for the needs of the populace. Perhaps the Ely *Isle de Vignes* is an association of ideas with the Bishop's garden sloping down to the Fleet in Holborn.

The Victoria County History of Hertfordshire has a few more references to vineyards of later dates that may conveniently be put in here. Richard Raynshaw in 1569 bequeathed a house called The Vine and some cottages in Spencer Street, St. Albans, for the benefit of the poor; the buildings being converted to almshouses.[82] At Bengeo Hall, near St. Leonards, Herts., are some cottages and a field called The Vineyard. A Mr. H. R. H. Gosselin-Grimshaw informed Mr. William Page (the editor of the Hertfordshire volumes) that the site had been used as a vineyard by Thomas Dimsdale, who had replanted the vines in 1767. At Thornley a field north of the road leading from the main thoroughfare to the church is called The Vineyard. In the north-west of the parish of Much Hadham there is a Vineyard Springs and a Vineyard Croft, dating from 1649. An inn at Hitchin, dating from 1617, is called The Vine, but, of course this no more implies that there was a vineyard there than does the name Mon Repos (in Laburnum Grove) mean that great peace reigns within that house. In fact the latter name would seem to imply the opposite, if we can believe our modern dramatists. As late as 1905 a Vineyard Field still existed at the Verulamium site, St Albans. A Mr. C. A. Ashdown excavated a Roman pavement from 'the north-west corner of Vineyard Field'.[2] It might well have been the site of a Roman villa with a vineyard attached to it. The place can still be seen. It is now a field about 300 yards south of the Roman theatre, on a gentle slope falling to the south-east and north. It is not an ideal site for a vineyard, but it is one that would do reasonably well and would provide good frost drainage.

A political and social vent which greatly affected our native vineyards occurred in about 1151 when the future Henry II fell in love with Eleanor of Aquitaine, wife of Louis VII of France (the lands she owned may have added to his passion). The Pope granted an annulment of her marriage with Louis on the grounds of incompatibility of temperament (which has a very modern ring) and in 1152 she married Henry. The wine-rich lands of Bordeaux thus came under the English crown and remained so until 1453. Fine claret wines were now freely available and Henry encouraged shipment from Bordeaux for several reasons. First, he was anxious to consolidate his hold on the lands he had acquired with his wife, and a prosperous wine industry would help with that. Henry saw himself more as a great continental prince than as King of England. Secondly, shipment of wine would call for more vessels, which could always be commandeered and used for naval purposes; and thirdly there was the revenue aspect. It was much easier to tax the wine arriving at a handful of ports than to keep tabs on a series of small vineyards all round the country, which, incidentally, with the present growth of English vineyards, is the position facing the Customs and Excise Officers today. A fourth and probably important reason was that other plants, particularly corn crops, were more profitable than wine and needed fewer skilled men to cultivate them. It was a period of great agricultural expansion and while better transport (particularly improved navigation) and better circulation of money (induced by increased gold and silver supplies) gave opportunities for specialization, in Britain the stress was more on increased production of food. The population was growing rapidly and exports were increasing.[100] Britain was on the way to becoming 'the granary of Europe'. But vines were not entirely neglected. For instance, William Somner (1598–1669) says that Henry de Eastry, Prior of Canterbury from 1285, indicated that both the church and abbey of St. Augustine's were well provided with vineyards and that they were also found at Colton, Berton, Chertham (now Chartham), Brook and Hollingbourne, all manors belonging to the abbey.[102]

The Manor of King's Langley, Hertfordshire, the remains of which are now underneath the Rudolf Steiner school there, was given to the church of St. Alban well before the Norman Conquest and was lost to the monastery in the disorderly times following the invasion. There seems to have been a vineyard there from an early period. The manor became a Royal Palace with an adjacent priory, and Queen Eleanor of Castile, consort of Edward I, went to live in the palace in 1276. An account roll of 1278 mentions 'a new start . . . a new garden with vines was laid out' and there are references to 'the gardeners' houses in the vineyards' and the 'long chamber towards the vines'.[74] In 1291/2 the sum of £66 13*s.* 4*d.* was paid to the mason Martin of Ray for making a wine cellar and it is this building that was recently (1970) excavated by David S. Neal. It was a fine structure, 74 feet long. In 1308 there was a reference to coopering tuns of wine, which suggests local production, for imported wine would come in its own casks, making extra barrels unnecessary. Additions were made to the cellar in 1388, including a window and a lock for the door. The palace seems to have gone by 1591 and the remains of the priory were demolished in 1691 by William Houlker. The church was pulled down in 1831 by a Mr. Betts who, however, did first make a plan of the building. The school now covers most of the site and there is no trace of the vineyard.

We now return to Pegge and continue with some more quotations from Somner. At Rochester and Sevenoaks there were two large pieces of ground 'now called *The Vine*' and there was Baron Sande's house in Hampshire bearing that name and still a showplace.[85] There was a vineyard at Halling in Kent which belonged to Bishop Haman. According to Lambarde, the bishop sent wine and grapes of his own growing to Edward, Prince of Wales (Edward II), when on a royal progress at Bockingfold, in the year 1303, when he was 19 years old.[56]

Pegge naturally has a look at William Camden (1551–1623), the historian and founder of antiquarian studies in Britain. Camden has two references to vineyards which are of interest. In his *Britannia* of 1586 he first of all notes the edict of Probus and Baron Sande's house, The Vine, to which reference has already been made, and then mentions that the vines were grown more for shade than fruit.[17] When Camden comes to Gloucestershire he quotes William of Malmesbury (died 1143) who obviously was enchanted by the place: 'The Vale of *Glocester* is so called from its chief city, the soil whereof yieldith variety of fruits and plants, and all sorts of grain; in some places by the natural richness of the ground, and in others by the dilegence of the Countryman; enough to excite the idlest person to take pains, when it repays his sweat with the increase of an hundredfold . . . There is no Province in *England* hath so many, or so good Vineyards as this County, either for fertility, or sweetness of the Grape. The wine whereof carrieth no unpleasant tartness, being not much inferior to the French in sweetness.'[59]

By the end of the sixteenth century these excellent vineyards seem to have gone. Camden did not believe that the soil had been worn out or that the climate had altered: '. . . we have no reason to admire that so many places in this County from their *Vines* are called Vineyards, since they formerly afforded plenty of Wine; and that they yield none now, is rather to be imputed to the sloth and unactiveness of the Inhabitants, than the Indisposition of the Climate.' All that remained of these old vineyards, said Camden, were the names, and Dr. Holland, the first translator (the original was in Latin) and editor, whose additions were collected in his successor's (Edmund Gibson's) 'Additions to Gloucestershire', gave two examples, one near Tewkesbury and the other 'on a rising hill by *Oversbridge*, near *Glocester*'.

Alas! Gone were those happy days when the working classes knew their place and worked. But they always were unsatisfactory. Lady Paston (1422–1509) wrote: 'Servants are not what they were wont to be'!

The 1695 edition of *Britannia* says there are no vineyards in Ireland: 'Vines grow here, but yield not so much benefit by their fruit as by their shade. For as soon as the sun is pass'd *Leo*, we have cold blasts in this country, and the afternoon heat in Autumn is too little, either in strength or continuance, here and in Britain, to ripen and concoct grapes to a full perfection.' The word 'Britain' above must refer to England and the statement thus appears to contradict William of Malmesbury, whom Camden seemed to believe. However, the Irish section was supplied by another pen than Camden's, that of Sir William Cox, Kt. Nevertheless Sir William seems to have known the sort of climate the vine needed, even if he did not think they could ripen in England. Camden also refers to an Essex Domesday Book vineyard: 'Not far from hence [Ashdown] lies Raleigh . . . there is in one park six Arpennies of vineyards, which if it takes well, yields twenty *Modii* of wine.' It seems that the *modius* was 36 wine gallons, equal to about 29 Imperial gallons. The yield was thus 120 gallons per acre, a very low figure by modern standards.

Pegge drew attention to Battely's Appendix and the prices paid for wine at an archbishop's enthronement, but there was some doubt about the particular archbishop and thus about the date.[7] It might have been Archbishop Warham's in 1504 or Archbishop Robert Winchelsea's in 1295—a difference of 209 years. Pegge thought it was for the former individual. The prices were:

Red wine	6 dolia at £4 per dolium
Claret	4 dolia price per dolium £73 0*s.* 4*d.*

(I cannot but think this is an error and that the above figure is at least the total for four dolia, i.e. £18 5*s.* 1*d.* per dolium.*)

White English wine, 1 dolium of selected		£3 6*s.* 0*d.*
White English wine for the kitchen, 1 dolium		£3 0*s.* 0*d.*
Malvesey	1 but	£4
Ossey	1 pipe	£3
Rhenish	2 almes	£1 6*s.* 0*d.*

What the capacity of these various measures was is difficult to estimate, but the figures do show relative prices—the best English wine was 16 per cent cheaper than the current run of imports.[82] In classical times the *dolium* was the largish vessel in which the wine was fermented.

Pegge has further references to prices. About 1260 a *dolium* of best wine cost 40*s.*, 2 marks or sometimes 20*s.* In 1420 red wine from Vasconia sold for 8 denarii and English wine for 6 denarii the flask, possibly then as uncertain a measure of capacity as the carafe in our restaurants today. Obviously English wine was about, but there was not much of it and it sold at a lower price than the imported vintages.

That the use of wine was not restricted to the church and upper classes may be seen from Chaucer's *Canterbury Tales.* His pilgrims, a cross-section of the community, took and enjoyed it. At the Tabard Inn 'Strong was the wyn and well we drynken wolde'. The Franklyn started on it early in the day, perhaps for breakfast, 'well loved he in the morwe a sope in wyn', and he appreciated quality too: 'His store of wyn was known in special'. It was probably imported. Lucky were the people who dined at his plenteous table.[21]

* The original reads: '*De vino clareto iv dol. prec. dol. lxxiii*[l] *iiii*[d]'. Perhaps the first 'l' was inserted in error for *libra* or £s and the price was really £23 0*s.* 4*d.* for 4 dolia, or £5 15*s.* 1*d.* per dolium.

According to Pegge the English vineyards were not confined solely to the south of the country. Dr. Plot said that the plant had been improved by Sir Henry Lyttleton at Over-Arley, Staffordshire, where, on a very favourable site, he made wine 'Altogether undistinguishable from the best French wines by the most judicious palates.'[87] Perhaps Sir Henry had developed a promising seedling suited to his conditions.

There was a vineyard at Darley Abbey, Derbyshire; and then Pegge brings up the subject of names, thinking that North and South Winfield and Wingerworth indicated that they were once vineyards.

There is also evidence of a vineyard having existed at Shrewsbury, on a south slope running down from the present hospital to the River Severn, a very favourable site. Pegge does not deal much with later periods but he does refer to a Captain Nicholas Toke who, according to Thomas Philipot (1659), had a vineyard at Godington, Great Chart, Kent, where he 'hath so industriously and elegantly improved our English wines, that the wine pressed and extracted out of their grapes seems not only to parallel, but almost to out-rival that of France.'[85, 86]

Pegge did not agree with the French historian Rapin de Thoyras who said that the repeal of Domitian's prohibition was of no great benefit to Britain. The contrary was the case: '. . . for the benefit was considerable questionless, although the British wines might not be of the richest and most generous kind, nor adequate in quantity to the consumption.'[109]

Pegge finished his article saying that the wine made by William Thorn in his abbey at Nordhome was 'agreable and greatly to be respected' and that the Duke of Norfolk made good wine, like Burgundy, at Arundel, Sussex.[108]

In *The Tempest* Shakespeare has an interesting line on vineyards.[97] Iris, in the masque scene, summons Ceres from her rural retreat, and describes it. Iris then refers to 'thy pole-clipt vineyard', which suggests that Shakespeare had actually seen vines pruned and trained to a pole—the goblet system, still in use today. Ceres later reports the bounty of the earth, including 'Vines, with clustering bunches growing', but this does not imply such direct observation as does the former line.

Judging from the literature, the seventeenth century saw a considerable revival of wine growing. There were at least fifteen books published on the subject or having considerable parts devoted to it, during the century, the best undoubtedly being that by John Rose.[95] It would be tedious to discuss all these works; most of them lament the decline in viticulture in England, show it is possible and press for its renewal. I shall consequently confine myself to discussing Rose's splendid work, as good a guide to the English vineyard today as when it was written.

John Rose became King Charles II's gardener at St. James's, after having served the Duchess of Somerset. The King was famous for his establishment of the 'plantations' in America and the fulsome introductory letter, obligatory in those days, carries this charming dedication to the King:

'The *Prince* of *Plants* to the *Prince* of *Planters*:

'This Royal Title, as Your Majesties (*sic*) great affection, and encouragement to all that is truely Magnificant and Emolumental in the *Culture* of *Trees* and *Fruits*.'

It has been suggested that the *Prince* of *Plants* was the pineapple. John Rose was the first man to get that plant to fruit in Britain and 'pine stoves' (greenhouses) soon became very popular, but the context of Rose's dedication leads me to think that Rose meant the vine in this case. Rose was encouraged to write his book by the gentle and cultivated John Evelyn (1620–1706), the diarist and author of *Sylva, or a Discourse of Forest-Trees*, etc. (1664), the family fortune (and hence the gentleness and culture) being derived from the selling of gunpowder alike to Royalists and Roundheads: business was business in those days too. The introduction was signed *Philoceps* and fairly obviously was by Evelyn, both because of its style, the references to *Sylva* and the importance of trees. Evelyn must also have been general editor of the book. In this same introduction *Philoceps* says he had talked with Rose and had turned over a mass of old books on the subject . . . 'But I do ingenuously profess that none of them have appeared to me more rational, and worthy of our imitation, than the short observations of Mr. Rose.' He then continued by saying that Rose's advice was the result of experience under practical English conditions and not just a following of overseas methods: 'Some *Monsieurs* new come over, who think we are as much oblig'd to follow their *Mode* of Gard'ning, as we do that of their garments, till we become in both ridiculous.'

Rose described six sorts of vine, mostly following Sir Hugh Platt's *Florae Paradisae*. Rose did however describe 'a new white grape found in the garden at St James's' which possibly was a successful seedling suitable for English conditions. The others were the 'White Muscadine', the Parsley grape, more a curiosity because of its leaf than a wine grape, the 'Muscadella' and the red and white 'Frontiniaq'.

Rose recommends a light sandy soil with a stony surface. It should be about two feet deep, have a bottom of chalk or gravel and should be free from springs of water. Blackberries growing profusely on it were a sign of its suitability for vines, but, of course the brambles all had to be removed before planting, a matter of some difficulty. Vines would grow well in rich loams, but they would be too vigorous and not produce as much or as good fruit. Dry spots were good for vines, once the plants were established in them. The site should have a slope to the south or south-west and be sheltered from the north and north-east by hills or woods.

The ground was to be ploughed in July. If it was grass, the turves were to be stripped and burned, and the ashes distributed over the site. Care was to be taken that the fires were not too fierce, because one did not want to burn the soil. Obviously Rose knew the value, on empirical lines, of organic matter in the soil.

The plants were to be set in rows three feet apart, and they were to run east and west, which is contrary to today's usual practice, but Rose gives his reasons. With close planting the east-west structure allowed the plants to benefit from the rising and setting sun, as the rows then would not shade each other. At midday the sun was high so that no shading of one row by the other occurred. The vines were planted in winter at a distance of two feet from one another in the rows, which gave 7,260 plants per acre, a large number. The vine seems to like crowded conditions, which is all very well if there is plenty of labour available. Each plant was staked and cut back to two or three eyes. The vineyard was kept well weeded. In January it was pruned, all growth being removed except the strongest cane, which was arched over and attached to the next shoot and another strong shoot, which was cut back to two buds, in order to provide wood for the next year. It was, in fact, Guyot pruning, long before that gentleman lived. The vineyard was dug in winter, but not too deeply in order not to destroy the surface roots.

In summer the growth was trimmed in order to limit the number of bunches per shoot. Rose thought one was enough. The shoots were to be broken off, not cut, as a broken wound healed more quickly. In late summer foliage was removed in order to expose the fruit to the sun. Maximum use had to be made of the sun, said Rose, and then the King, who 'may not of had much opinion of English wines up to the present', would change his mind. From now on gentlemen would find that 'that precious *Liquor* may haply once again recover its just estimation'. Finally Rose offers to supply plants of suitable wine varieties.

That the book was a success may be seen not only from the numerous editions it went to, but also by the fact that other authors kept referring to it, though whether it led to any extensive planting up is not known. In 1677 Henry Oldenburg, secretary to the Royal Society, published a book *Nurseries, Orchards, Profitable Gardens and Vineyards Encouraged*, in which he highly praised John Rose, but lamented that not much was being done: 'But yet I have much to say for the Wine of the Grape, though with some disparagement to our own Country-men, who have done so little for it, after they have such bright Instructions, and such encouragements from Mr. *Evelyn* and Mr. *Rose* . . .' It must also be noted that Evelyn, though an enthusiast for English wines, kept an open mind on the subject. In 1655, *before* Rose's book, he wrote in his diary: '. . . to see Colonel Blount's subterranean warren and drunk of the wine of his vineyard, which was good for little.'[32] Presumably Rose changed all that.

Although Rose was remarkably sound on the growing of grapes, he did not have the same skill in actually making the wine. In the seventeenth century there were two advances in this art. The first was Chaptalisation—years before Chaptal (1756–1832) was born—the adding of sugar to improve musts.[52] Comparatively cheap sugar was now arriving from the West Indies. The second was the use of a primitive hydrometer to estimate the strength of wine musts—a vital aspect of wine making today. Sir Kenelm Digby (1669) told his readers to add honey to a must until 'a fresh egg floats to the depth of twopence'.[25] The more sugar in the must the higher the egg would float. Much depended, of course, on the freshness of the egg. A not-so-fresh egg, containing more air than a new-laid egg, would float higher and give a false reading. A fresh egg, an almost unknown luxury today, was a common object in Sir Kenelm's times and was a good guide to the strength of a must.

English vineyards continued to arouse interest in the eighteenth century and the flow of books continued. There were at least eighteen of them dealing exclusively or in a large part with English vineyards during that period. Once again it would be tedious to discuss them all, so I shall just go into the general trend and describe a famous vineyard—Pain's Hill—from Sir Edward Barry's book, though it is mentioned in several others as well, and cast an eye on Miller.

In spite of Oldenburg's complaint mentioned above, the planting of vineyards does seem to have continued, but to some extent the discovery of the hot-house diverted interest from the open-air vineyard and a number of trials made with unsuitable grape varieties and at bad sites received considerable publicity and discouraged all but the most determined growers. The three great eighteenth-century vine authorities were Phillip Miller, William Speechley and Sir Edward Barry.

Miller (1691–1771) was gardener to the Apothecaries, which garden still exists on Chelsea Embankment, London. He produced *The Gardeners Dictionary*, a famous work which went to eight editions in his lifetime. That he was interested in viticulture may be seen from the facts that the last word of the rather prolix title is 'Vineyard' in large type and that the article on *Vitis* runs to 118 folio columns, say 65,000 words—a respectable book, in short. Fourteen of these columns are on English vineyards.[67]

Miller, in contrast to Rose, favoured wide spacing, the rows to be ten feet apart and the plants at six-feet distance in the rows—726 plants per acre. Miller maintained that wide planting enabled air to circulate and dry out harmful fogs and dews. It should be noted that this was before the two bad mildews arrived from America, but the *Botrytis* fungus, now causing trouble to our vineyards, could have been present.

Miller next discusses the varieties that should be planted and says that much of the failure of recently planted vineyards was due to the cultivation of dessert varieties which were of little use for wine. He describes the usual planting and cultivation methods and says the rows should run south-east to north-west. Each vine was to have a stake, and training and pruning are once again basically the Guyot system. He advocates what sounds like exceptionally early 'winter' pruning, at Michaelmas—29th September—but this was the 'old style' date. The calendar was altered on 3rd September 1752, the following day being called 15th September, leading to the famous 'Give us back our eleven days' riots. Thus the true date of Miller's Michaelmas was our 10th October. Even so it is very early for winter pruning judging by today's standards—the middle of harvesting in most modern British vineyards.

Another point in Miller's instructions is that he strongly advises against stripping off the leaves to expose the fruit to the sun: 'And here I can't help taking notice of the absurd Practice of those who pull off their Leaves from their *Vines*, which are placed near the Fruit, in order to let in the Rays of the Sun to ripen them; not considering how much they expose their Fruits to the cold Dews, . . . which, being imbibed by the Fruit, do greatly retard them.' Today we are still not quite sure whether the practice is advantageous or not. The matter will be discussed below.

Miller also imported American vines to England in the early eighteenth century and it is fortunate that he did not bring into the country the phylloxera and the two dangerous mildews.

In addition to the article on the vine Miller has another long one on wine and the wine-press (51 folio columns) with two large plates showing the construction of a complex wine-press.

Miller obviously had collected a great deal of empirical knowledge on the making of wine. He advises the use of Sir Kenelm's 'fresh egg' to regulate the addition of honey or sugar to musts, though he does not actually refer to the knight. He speculates at length on the nature of fermentation and the 'vinous spirit' it produces and does not use the word 'alcohol'. He is also intrigued by the bubbles rising in the must which 'burst with an audible noise'. He realizes the value of sulphur fumes as a disinfectant and at one point quotes figures from Stephen Hales on the amount of bad air produced; it could make people ill, he says. Obviously it was carbon dioxide. Another interesting feature is something like an air-lock in barrels in order to prevent wine souring, and there is a great deal on additions to wine in order to overcome troubles in the wine itself—sourness, bad smells, ropiness and so forth. Many of these substances are fining agents, such as egg-white, milk, isinglass, ashes, calcined flint and flour. Others, such as powdered alabaster, marble and chalk would both clarify the wine and reduce acidity. To increase acidity a gill of 'oil of sulphur', presumably sulphuric acid, can be added to a barrel of wine! Colour could be restored to claret by adding the juice from boiled beetroots or an extract of elderberries. Flavours and bouquet could be obtained by adding extracts of beech bark, cloves, cinnamon, nutmeg, elder flowers, ginger, orris root and saltpetre. At the same time Miller points out that many of these troubles can be avoided if the wine-making apparatus is kept clean and exposed to the fumes of burning sulphur before use, and he describes how to make sulphur wicks. Miller's two plates show two substantial wine-presses based on the principle of the cider press; the grapes were packed into cloths and piled on the press bed; between each cloth was a frame of wooden battens. The cloths were then squashed by a system of levers and ropes. The worm of the press was of wood, usually oak, thirteen inches in diameter.

The vineyard at Pain's Hill, Cobham, Surrey, is famous. It was planted by the Honourable Charles Hamilton in the early eighteenth century and the best account of it and its wine is that given by Sir Edward Barry, a man with great faith in the methods advocated by Phillip Miller.[5] The Honourable Charles Hamilton was the youngest son of the sixth Earl of Abercorn. The young man laid out a natural landscape garden on the slopes of St. George's Hill, above the Mole Valley, and that is where he established his vineyard.

Hamilton used two kinds of grape, the 'Auvernat' and 'Black cluster', this last being a synonym for the 'Pinot Noir' of the Champagne, also known as 'Noirien' and 'Beaunois'. It is a quality grape and not a heavy cropper. What the former was it is difficult to say; presumably it was a variety from Auvergne, not a famous wine province.

The two grapes used were both black and the first wine made was not very good and so Hamilton turned to making whites. 'This essay did not answer; the Wine was so very harsh and austere, that I despaired of ever making red Wine fit to drink; but through that harshness I perceived a flavour something like that of some small *French* white Wines, which made me hope I should succeed better with white Wine. That experiment succeeded far beyond my most sanguine expectations; for the very first year I made white Wine, it nearly resembled the flavour of *Champaign*; and in two or three years more, as the Vines grew stronger, to my great amazement, my wine had a finer flavour than the best *Champaign* I ever tasted; the first running was as clear as spirits, the second running was *oeil de Perdrix*, and both of them sparkled and creamed in the glass like *Champaign*. It would be endless to mention how many good judges of Wine were deceived by my Wine, and thought it superior to any *Champaign* they ever drank; even the duke *de Mirepoix* preferred it to any other Wine; but such is the prejudice of most people against anything of *English* growth, I generally found it most prudent not to declare where it grew, till after they had passed their verdict upon it. The surest proof I can give of its excellence is, that I have sold it to Wine merchants for fifty guineas a hogshead; and one Wine merchant, to whom I sold five hundred pounds worth at one time, assured me, he sold some of the best of it from *7s. 6d.* to *10s. 6d.* per bottle.'

We have no indication of the size of the Pain's Hill vineyard, nor can we examine the site, for some high-rise flats are the new crop there, but the scale of operations can be judged from some of the facts mentioned above. The merchant who spent £500 on that wine must have bought at least 10 hogsheads, say 460 gallons of wine, the produce of 1½ acres. Presumably he was not the only outlet; the proprietor himself obviously enjoyed and used the wine and there may well have been other buyers. The merchant too must have done well out of the transaction. He paid £52 10*s*. for the hogshead, out of which he would have got at least 480 bottles, which, at 7*s* 6*d*. per bottle, was £180. Of course he had to provide bottles, corks, labour and transport, the last item being most expensive.

Hamilton used a strange method to make his wine. He let the grapes hang as long as possible, then picked them and brought them to the winery in small quantities. The fruit was picked off the stalks, all mouldy berries discarded and the sound crop put into the press. The free run, the juice that ran out of its own accord before any pressure was applied, the first pressing and part of the second ran white and were kept separate from the subsequent runs which were reddish. The must was put into hogsheads and closely bunged. Fermentation soon started and every effort was made to contain the gas within the cask. The pressure developed must have been enormous, and very dangerous too: 'In a few hours one could hear the fermentation begin, which would soon burst the casks, if not guarded against, by hooping them strongly with iron, and securing them in strong wooden frames, and the heads with wedges; in the height of the fermentation I have frequently seen the Wine oozing through the pores of the staves.'

One cannot help feeling that Hamilton must have been as dangerous as Guy Fawkes, but he says nothing about any explosions. The casks were left in a cold barn for the winter, and in cold weather were racked into clean casks. Fining was carried out only if needed, isinglass being used. Bottling was done at the end of March. In spite of Hamilton's precautions much of the gas must have escaped. However, bottling in cold weather did ensure that some gas from the original fermentation was still present and as the temperature rose a residue of sugar in the bottles would start to ferment again, producing more in-bottle gas. Hamilton says nothing about *dégorgement*—getting the yeast deposit out of the bottles. Perhaps in those days they did not worry about such details. They drank it young too: 'and in about six weeks more [the wine] would be in perfect order for drinking, and would be in their prime for above one year; but the second year the flavour and sweetness would abate, and would gradually decline, till at last it lost all flavour and sweetness; and some I kept sixteen years became so like *Old Hock*, that it might pass for such to one who was not a perfect connoisseur.'

The wine obviously had lost all its gas after that long wait, but it is strange that Hamilton should say it had lost all flavour, because one would have expected that to develop with the passage of time. Had Hamilton's palate lost its skill in the sixteen-year interval? 'The only art I ever used', he says, 'was to put three pounds of white sugar to a hogshead of must because of the rage for sweet Champagne.' This is hardly Chaptalisation—the quantity is so small and as it made the wine sweet it suggests two things: first, that the Pain's Hill must was high in sugar, and secondly that the local strain of yeast was intolerant of alcohol and stopped fermenting at a low point, say 10 per cent.

When this successful vineyard faded away I have been unable to ascertain: it might have been connected with a patent, granted to the Reverend Phillip Le Brocq in 1785, for an elaborate method of planting and training vines.[13] It was also described in a book published that same year, wherein the author threatened dire punishment on all who used his system without payment of a fee.[14] The method was too involved to be of use, but many people must have hesitated to plant up for fear of infringing the patent. However, a Monsieur Vispré, one is glad to see, defied Le Brocq in a book published the same year.[113] Vispré thought wine growing would flourish in England were it not for Le Brocq's patent, which not only advocated wrong methods but also was so vague that people were afraid of planting up in any style, through fear of infringement. However, Le Brocq did realize the importance of concentrating the sun on this crop.

A most important event at the end of the century was the rediscovery of the cork.[52] Vintage wines developing in bottle were now possible, and ordinary wines, such as most of those from English vineyards, could be kept more or less unaltered in bottle for a long time.

In the nineteenth century sugar started coming into Britain from the West Indies in ever-increasing quantities. Commercial wine makers started using it in such big amounts that they spoiled a promising industry. Instead of trying to establish English wines on their own merits the would-be vintners started to imitate foreign vintages. The Napoleonic Wars theoretically should have given the English a fresh start, because the war was supposed to have cut off supplies from France. In fact the prohibition was avoided; wine was shipped via neutral ports, as diplomats' baggage, and was also smuggled in large amounts, particularly from Bordeaux with its strong English connections, all adding greatly to the cost.

J. Davies's *Inkeeper's and Butler's Guide, or a Directory of British Wines* had gone to six editions* by 1808 and gave directions for making numerous imitations. A similar publication was W. H. Roberts's *The British Winemaker and Domestic Brewer*, which had five editions between 1835 and 1849, the first, third, fourth and fifth being now to be found in the British Museum. That book was important because it gave full and accurate instructions for the use of the hydrometer. The chief characteristics of its recipes were

* The fact that a book is marked '6th edition' does not necessarily mean it is the sixth *edition*. Printers and publishers in those days might mark a reprint so, to show how important and popular the publication was and thus stimulate sales.

the great strength and sweetness of the resulting liquids. For instance, gooseberry wine 'similar to Champagne' was made by soaking gooseberries two-thirds ripe in water, pressing and making up the must to a gravity of 1.10. This is equivalent to 3 lb. of sugar per gallon which, if it all fermented out, would give an alcohol content of 17·7 per cent, beyond the powers of most yeasts (though in 1945 I made some Kentish wine that went to 18 per cent alcohol—it was a wonderful vintage year), so presumably some remained behind, giving a sweet wine. Not content with this, the author advised adding 2 quarts of brandy per 5 gallons of wine when fermentation was over, which would add another 5 per cent of alcohol! The resulting liquor would be not in the least like champagne, but a strong, sweet cordial.

No wonder 'British Wines' got a bad reputation and were much mocked; one has only to think of Mr. Pooter's standby for all crises and celebrations—a glass of *Jackson Frères* champagne from the grocer's, at 3*s.* 6*d.* per bottle.[42] Strangely enough there is a genuine Champagne house at Rheims called *Jacquesson*. Whether this is a question of nature imitating art or of Mr. Pooter's grocer's promotional skill is difficult to say.

A few British vineyards persisted, the most famous being that of the Marquis of Bute, planted in 1875 at Castle Coch, Glamorganshire, and continued at Swanbridge. Lord Bute's gardener, a Mr. Pettigrew, ran the vineyard and after twenty years opined that wine growing could be made to pay in Britain. And this was in spite of the troubles from America: three pests, only one of which was serious in Britain in Victorian times. More is said about these below.

1881, the year of the great snowstorm, produced a good vintage, most of which was sold for 5*s.* per bottle. Some of it resold at auction a year later, by one of the purchasers, fetched £5 15*s.* per dozen, which was still not as high a price as that obtained by Charles Hamilton more than a century earlier.

Many country houses, cottages and even some town dwellings in Victorian times had vines on sunny walls from which, in the early part of the century, the owners made wine. The troubles from America, already mentioned, discouraged this. The three pests were the powdery mildew (*Uncinula necator*), the phylloxera (*Phylloxera vastatrix*) and the downy mildew (*Plasmopara viticola*). As far as British vines went the powdery mildew was the worst. It was first found at Margate in a glasshouse by a Mr. Tucker, gardener to Sir John Slater, in 1845.[10] A fungus disease, it was named *Oidium Tuckeri* by the Reverend M. J. Berkeley; he saw only the vegetative stages and so named it wrongly. Three years after his death (1889) the sexual forms were discovered and the fungus found to belong to the genus *Uncinula* Lév. and given the specific name of *necator* Burr. The white powdery mildew spread over the leaves and fruit, slowly and surely destroying them. It spread rapidly and soon was devastating indoor and outdoor vines all over Britain and the rest of Europe; it still is a pest (see page 101).

The cure was simple and found fairly quickly—spraying or dusting with sulphur, a treatment that became a routine operation in commercial vineyards. It was, however, a different matter with householders and gardeners. Treatment had to be given at the right time, repeated at proper intervals and done systematically. Neglected vines on a garden wall, which once had produced a fine crop of grapes, were now attacked. Their wretched appearance gave rise to the view that the climate was not suitable for vines and discouraged grape-growing experiments in Britain. In 1860 the 'penal duties' on French wines—1*s*. per bottle—were reduced by Mr. Gladstone to 2*d*. per bottle, making imports yet cheaper and further discouraging home production.

The phylloxera, an insect from America which slowly and surely killed the European vine, strangely enough was first found in Europe at Hammersmith, in 1863. It caused untold damage in continental Europe, but none to speak of in Britain, where it has never become established—fortunately.[80]

The downy mildew was another fungus from America: first found in France in 1878, it had been described in America as early as 1834. M. Cornu, who did so much to overcome the phylloxera, as early as 1873 warned the vignerons about introducing this disease along with the American *Vitis* wood imported for phylloxera control, but little notice was taken of him. Soon the downy mildew was attacking vines all over continental Europe. Once again control measures were comparatively easy, spraying with 'Bordeaux mixture'—copper sulphate and lime. The disease did not do much damage to British vines and was only recorded in England for the first time in 1894 and then not again until 1926, but it is now fairly well established (see page 104).

The twentieth century opened with the Castle Coch and Swanbridge vineyards flourishing and being extended, but the cottage vines were succumbing more and more to the powdery mildew. In the late nineteenth and early twentieth centuries vine-houses flourished. Labour and coal were cheap and magnificent bunches of table grapes were produced in the hot-houses by skilled gardeners who pruned and tied the vine rods and thinned the bunches with special grape-thinning scissors. Big commercial vineries arose and the luxurious bunches produced were even exported to America. The fruit was not suitable for wine and in any case would have been far too expensive for that purpose. Thus the impression grew that in Britain vines could only be grown under glass. The Marquis of Bute's vineyards were far away and not many people knew of them, in spite of their being described in George Bunyard's book (1904).[15] They disappeared during the First World War.

To close this chapter we might once more consider the question of place names. There is a Vineyard Farm at Claverton, Somerset, and a field there is called Vineyard Butts. It obviously was a vineyard and probably fairly recently, this century at any rate, because it is a sloping field in a bank facing south and the terraces can still be seen. I went to look at it about 1950 and the farmer told me that when she took it over 'before the war' there were still a few vines around, but she added 'we soon got rid of them'. A pity, because it would have been interesting to see what sort of vine was being grown. There were also the remains of a big press in the outbuildings, which of course, could have been a cider-press as well. It was an ideal site and I was told recently that it is to be replanted with vines.

John Field in a recent book on farm and field names has found numerous references to vines, such as The Vine(s), Vine Farm, Viney Mead, Vyneheye (Old English, 1294 (ge)hoeg—'land on or near which grapes were grown'), The Vinery and The Vineyard.[33] The last name is quite common and shows a widespread distribution over England; for instance, at Hale, Cheshire; Buckland, Devon; Brailsford, Derbyshire; Arne and Corfe Castle, Dorset; Clifton, Fairford and Staunton, Gloucestershire; Preston Montford, Shropshire; Mytholmroyd, West Riding, Yorkshire; Ely, Cambridge, etc.

3
The vineyard revival

In the inter-war years a kind of revival started; small clubs were established making wines from flowers and fruits, rhubarb, elder, gooseberries and so on, even, at times 'grape wine'. I myself was among these people, but I began to think more about real wine because I liked it and my work on pest control experiments took me to the Champagne, to work in the vineyards on rooting hormones and the destruction of the grape caterpillars (*Cochylis* and *Eudemis*). Fortunately these pests, like the phylloxera, are not endemic in England. While in that part of France I noted that the climate was very similar to that of southern England, except that our winters were considerably milder. Furthermore, in Kent there were many old houses with substantial but nearly dead vines on them. Mereworth Castle (a Palladian building, not in the least like a castle) was one, and a cottage by the main road at Wrotham was another. It was not the climate or soil exhaustion that was killing these vines but the powdery mildew, and if I couldn't control that, I thought, I was not much good as a pest-control expert. In 1936 I struck cuttings from vines growing in the neighbourhood, and obtained plants from local nurserymen (who tended to recommend 'Black Hamburg', useless for outdoor purposes) and the following year planted vines on the sunny walls of the house and garden and also in 'high culture' wires across the garden.[78] I kept down the mildew with sulphur dustings. The first vintage was in 1940. One of my vines was *Vitis purpurea*, a grape with highly coloured flesh and skins: as I was making red wine, a small quantity of *purpurea* greatly helped the colour.

Production was only on a garden scale—some sixty gallons a year (a bottle a day)—but much useful experience was obtained. For instance it soon became obvious that it was better to make white wine rather than red, though some good red wine was made in the three exceptional years of 1945, 1947 and 1949. My main variety at that time was 'Brant', not really one of the best; it has a suspicion of *labrusca* flavour in it and ripens a little late.

Soon after the war Edward Hyams planted the first true vineyard of the revival; it was near Canterbury, Kent. Later he set another out at Ashburton, Devon. In 1949 Edward Hyams published the first book of the wine-growing revival, which did much to draw attention to the possibilities inherent in the outdoor grape.[48] He also discovered that the black grape on the cottage by the main road at Wrotham, Kent, was a strain, possibly a bud sport, of 'Pinot Noir' very suitable for England, and he obtained cuttings from it, fortunately, for the cottage and the vine have now been destroyed in the name of road improvement. The BBC and the weeklies took up the subject and further books and articles followed.[27, 49, 77, 122] The public became more and more interested.

Nurserymen began to sense the possibilities of the new market for outdoor vines. Varieties 'suitable for the open vineyard' were soon on offer, but unfortunately little or no testing was done as to their suitability for our climate. In some cases large stocks of what were supposed to be early varieties were simply purchased on the continent and resold in Britain. Foreign nurserymen sometimes delivered wrongly named material, probably thinking that in England—land of perpetual fog—the vines would not grow and, since wood or rooted *vinifera* cuttings were ordered, if they did grow the phylloxera would soon kill them, so it did not much matter what was sent. A fool and his money were soon parted.

Some pioneers were discouraged by having planted wrongly named material, but, nevertheless, some good cultivars were set out, showing what could be done. This is where Mr. Barrington Brock became important. Entirely at his own expense, in 1946, he established the Viticultural Research Station, at Oxted, Surrey, and did a great deal for the modern revival of vineyards in Britain. The information was given to the world by the Reports he issued and the 'Open Days' he held at the station.[12] It is slightly ironical that it was his namesake, two centuries earlier, who did so much harm to English vineyards (see page 43); amends have now been made.

One of Mr. Barrington Brock's first tasks was to sort out the vexed question of varieties and he soon found there was a vast amount of false information in the literature. One of the difficulties of all forms of horticultural investigation is that of varietal synonyms and misnaming, being particularly true of vines. For instance, Lorenzo R. Badell, who compiled a list of vine varieties in 1952, enters thirty-two synonyms for the 'Pinot Noir' and the different kinds of 'Gamay' fill four pages, Gamay Beaujolais alone having seventy-three other names.[3] One can see the sort of difficulties Mr. Barrington Brock faced. It is extremely annoying, after patiently waiting at least three years, to find that, say, 'Clifton's Wonder' is exactly the same as 'Frederick's Early' and that neither of them is much use, but the value of such information to the grower is enormous. The same sort of thing happened with apple rootstocks. When the East Malling Research Station started to examine 'Paradise' stocks they isolated at least eighteen different kinds, some giving a tiny dwarf tree and others a giant. The work took some ten years and the modern, efficient apple industry would never have been started without it. Nor would the modern revival of vineyards in England have got off the ground so quickly without the Oxted station. Its reports formed a guide to the growing band of would-be vignerons, both amateurs making wine for their own use and larger scale operators growing it for sale.

The Viticultural Research Station proved to be immensely popular, but also very expensive for an individual to run. Postal enquiries reached 2,000 a year; all were answered, though very few people bothered to send a stamped, addressed envelope. Vines of promising varieties were sold, but never covered the cost of running the station. Applications for supporting grants to help this station, which after all showed all the signs of being able to start a promising new industry, never succeeded, the Agricultural Research Council taking the prize for banality. They thought the idea very interesting but could not grant any money unless Mr. Barrington Brock had first shown the scheme to be successful! It reminds one of Edmund Burke: 'The only criterion of vulgar judgements—success.'[16]

The strangeness of the idea to the official mind was always evident: at first the losses on this valuable research station could be set against the proprietor's personal tax, but after five years' work, when most promising results were coming to hand, this was stopped because, the authorities said, Mr. Barrington Brock had not proved the exercise was ever likely to be a successful farming enterprise. Even so our pioneer bravely carried on until 1970 when he closed the station. The research he had set out to do had been completed, costs were soaring and the subject was now being taken seriously in many official places. Mr. Barrington Brock does still grow his approved varieties, which is fortunate, because it is a standard collection of certified names. In years to come nurserymen may well get the varieties muddled up and then the Oxted collection will be a reference point. How fortunate that the Lawes brothers, who started the famous Rothamsted Research Station in 1834, did not face the same sort of trouble.

The Customs and Excise were also worried by the strangeness of the project. The amateur may make as much wine as he likes for his own use; he may give it away but may not sell it. Nor may he distil it in any shape or form, neither for home use nor for sale (see page 146). Mr. Barrington Brock thought that some of his experimental wines would make good brandy and applied for a licence to distil small amounts for experimental tastings. Customs and Excise had never issued a brandy licence before and, over a year, made every conceivable objection to the proposal. When Mr. Barrington Brock said he was fed up with waiting and was going to send all the correspondence to his M.P. they (Customs and Excise) granted him a licence, hedged around with many conditions, one of which was that none of the spirit must be drunk but all destroyed immediately after production! However, fortunately in practice the inspectors were reasonable about quantities used for testing. Some good brandy was made.

The importance of the Oxted work can be seen from the fact that two English research stations now have experimental vineyards—Long Ashton, Somerset, and Wye College, Kent, while the Ministry of Agriculture's advisory service will deal with certain enquiries.

The first professional of the revival was Major-General Sir Guy Salisbury-Jones who planted some three acres of vines at Hambledown, Hampshire, in 1952 and was soon producing some fine wine. Another important pioneer was Mr. J. L. Ward, living on the Kent–Sussex border. He was interested in cider and country wines and made a discovery which nearly drove the government mad. This was that while beer and wines paid excise duties, cider was tax free and there was nothing to prevent apple juice being

Chaptalised to any desired degree. Mr. Ward's company, the Merrydown Wine Co. of Horam, Heathfield, Sussex, marketed a very strong cider which was in effect a very agreeable apple wine —clear, full flavoured and pleasant, with about 12 per cent alcohol, and cheap because, being 'cider', it paid no excise duty. Special legislation had to be introduced in 1956, when cider with more than 15 degree proof spirit (8·55 per cent alcohol by volume) paid excise duty as wine.

The Merrydown company also planted a vineyard and made wines from currants, gooseberries and other fruits, either for their own account or for growers. Soon they were making wine (real wine—'grape wine') from their own grapes or purchased English grapes. They would also make wine for growers for an agreed service fee and have helped many beginners in this way.

Another important early pioneer was Mrs. Gore-Browne, at Beaulieu, Hampshire. She had the courage to start on a big scale—seven acres—and to install a winery with up-to-date and first-class equipment which resulted in very good wines.

English and Welsh vineyards began to expand and it was soon obvious that there were three sides to the development of the new industry: 1. Growing the right kind of grapes; 2. Making the wine correctly; 3. Removing the prejudices against English wines, the last a matter that troubled the Honourable Charles Hamilton two centuries previously (see page 41).

Vine growers seem to be friendly people and help each other a great deal, so that, apart from Mr. Barrington Brock's welcome Reports and 'open-day' discussions at Oxted, a great deal of useful information was exchanged among them. At the Long Ashton Research Station Mr. A. Pollard contributed much on the making of the wine, and later on Mr. Anton Massel established the Oenological Research Station, and a vineyard, at Ockley, Surrey, where sound professional advice on wine making can be obtained.

As to removing the prejudice, much help was given by the simple existence of Sir Guy Salisbury-Jones and his vineyard, let alone the excellent wine he makes. He was Marshal of the Diplomatic Corps in London and if he could make and drink and sell his own wine surely that was good enough.

The number of English vignerons slowly grew and in 1967 the English Vineyards Association was founded, with Sir Guy Salisbury-Jones as president and Mr. J. L. Ward as chairman. This important body flourishes today. Its present secretary is Mrs. I. M. Barrett who, with her husband, has ten acres of vines at Felsted, Essex. In 1974 they even made some excellent 'ice wine' (see page 127), the grapes being picked at 2 a.m. on a freezing November morning and processed at once.

English wines compare very favourably with those of Germany and northern France. Quality very often arises in crops growing under marginal conditions, as has been mentioned on page 17, and the would-be vigneron is well advised not to choose too rich a soil or to manure heavily. By such restraint he will produce a wine second to none.

The English wine growers are devoted to their task: there are about 385 in the Association with over 500 acres planted. In blind tastings the quality has compared favourably with Continental wines from the same varieties of grape.

4

Some problems of British vineyards

The problems of British vineyards fall into five categories, namely those connected with sites, varieties, stocks, pruning and pest control, each of which will be dealt with in turn. A great deal of information on all subjects is still needed and a point to remember is that the vine is a complex plant and that results obtained abroad, even in the northern European area, will not necessarily be of the same value in Britain. Also, the vineyard revival, using the new varieties and methods, is so comparatively recent that I can do no more than point out the problems and the various solutions suggested and used without being able to indicate which is the best. Everyone tends to have a favourite method. Alexander Pope expressed it neatly:

> *''Tis with our judgements as our watches, none*
> *Go just alike, yet each believes his own.'*[88]

Sites

The vine to ripen its fruit well needs sunlight and heat, and since the site must be selected to provide both, a south-facing position is demanded. Light is necessary for photosynthesis, actually making the sugars and other substances we need for the wine. The process also needs heat, but what constitute the optimum requirements of these two factors we do not yet exactly know. Will more light compensate for a little less heat? The farther north one goes the more potential sun there is in summer—that is, the duration between sunrise and sunset; I am not considering cloud or actual hours of sunshine at the moment.

SCOTLAND

We might well consider if a vineyard on the west coast of Scotland would ripen grapes as well as one on the south coast of England, that part of Scotland being under the influence of the Gulf Stream.

Let us examine the theoretical difference in hours between sunrise and sunset in, say, Stornoway, South Uist, Scotland, and Sandown, Isle of Wight, whose latitudes respectively are 58° 13′ North and 50° 36′ North, a difference of 7° 37′ latitude. These are near enough for our purpose to 58° N and 50° N, figures for which there are entries in the Nautical Almanac.[73]

Differences in intervals between sunrise and sunset
Midsummer in two northern latitudes

Latitude	*Sunrise June 21 (hours)*	*Sunset June 21 (hours)*	*Interval (hours)*	*Sum of Intervals Spring to Autumn Equinoxes (hours)*
58° N	2·93	21·12	18·19	1,655·29
50° N	3·83	20·22	16·39	1,491·49
Difference			1·82	163·80

(The entries above are in hours and decimals of an hour)

At first sight 164 more hours of possible sunshine in the more northerly latitude looks like a considerable difference, but it is only 10 per cent more than the more southerly potential. On the other hand the northern position has a much longer twilight; in fact in Uist at midsummer it hardly gets dark at all.

The temperature figures—means of thirty and twenty-six years —are given below. They are the sum of the means between mean maximum and mean minimum for the two periods stated, divided respectively by six and four (six months being the period between the equinoxes and four months—June to September—the main growing period).

Place	Mean temperature *April to September*	*June to September*
	°C	°C
Sandown, I.O.W.	15·2	15·475
Stornoway, Scotland	11·45	12·075
Difference	3·75	3·4

There is thus a considerable difference in mean temperature between the two places, Sandown being 33–28 per cent warmer than

Stornoway. We would hardly expect a 10 per cent advantage in possible light to make up for the 30 per cent loss of heat. Moreover, it is only a potential gain in summer sunshine. The sunshine records show that whereas the more southerly site gets from 38–51 per cent of the potential, the more northerly one only receives 24–38 per cent of it, still further disadvantaging the latter. However, to counterbalance this there is also some radiation derived from the sky alone which can be considerable in northern latitudes, 80 per cent of the total radiation received in winter and 25 per cent in summer.[89] Even so this would only increase the Stornoway summer figure to 12·5 per cent more potential radiation than Sandown, still not enough to compensate for the loss of heat. We can only conclude that in spite of the influence of the Gulf Stream, north-west Scotland is not suitable for outdoor vines.

Site orientation

We next have to consider questions of site orientation, direction of rows, planting distances and micro-climate.

MOVEMENT OF THE SUN

To discuss site and orientation we need to give some theoretical consideration to the sun's movements. There are two aspects to deal with. First, the depth of atmosphere through which the sun's rays must travel to reach us, and second what happens to any given area of a bundle of rays entering our atmosphere when it reaches the earth. Now for the first aspect. Insolation depends on the incident angle of the sun's rays—the angle at which they strike the vineyard. The atmosphere absorbs and scatters the sun's rays to the extent of about half their strength before they reach the ground, which is fortunate for us, for otherwise we should be killed. The less atmosphere the rays have to pass through, the stronger the insolation received will be. The sun is so far away that to all intents and purposes the rays are parallel. An acre of level land in the tropics at certain dates at midday will get an acre column of sunlight impinging directly on it, the rays having the shortest passage through the air. As the sun sinks (I know it is the earth that moves, but, mindful of Galileo's fate, it is more convenient to think of the sun as moving!) the acre column of air has to pass through more air to reach the ground and consequently more insolation is absorbed.

AREA ILLUMINATED

The second aspect is that as the sun sinks the acre column entering the atmosphere is spread over a greater area. When the angle of incidence to the horizon from level ground is 45° the acre column hits an area of 1·5 acres and at 22·5° over the horizon the acre column of sunshine is spread over 2·5 acres. It can be seen that there is a considerable falling off in the strength of a low sun; this is the reason for the blindingly obvious fact that the tropics are hotter than the temperate zone!

Nowhere in Britain (in fact nowhere outside the tropics) is the sun ever directly overhead, but the slope of a piece of land can have the effect of the sun's being overhead as regards the second aspect mentioned above (it does not affect the first aspect—the length of the journey through the atmosphere).

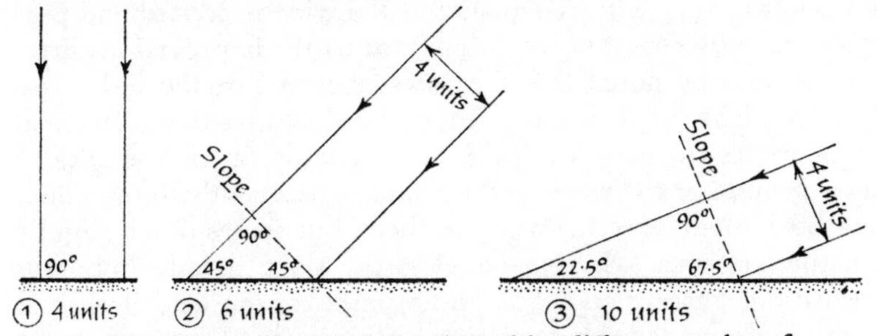

Fig. 1. Varying areas of land illuminated by different angles of incidence of the same bundle of sun's rays.

(1) A column of sunlight striking a level vineyard at 90° (i.e. sun directly overhead) illuminates 4 units of land.

(2) The same column of sunlight striking level ground at 45° is spread over 6 units of land. If the ground slopes 45° the column will illuminate 4 units again.

(3) The same column of sunlight striking the level ground at 22·5° is spread over 10 units of ground. If the ground (say a cliff face) is at 67·5° the column illuminates 4 units again. Note that in the cases of the slopes in 2 and 3 that though these slopes restore the direct impact of the sunlight the column will still have travelled through more atmosphere and consequently will be that much weaker.

SLOPE AND SUN INCIDENCE AT 51°N (UCKFIELD, SUSSEX), NOON, 21ST JUNE

Under the above conditions a vineyard facing south with a slope of 27·5° will have the sun at right-angles to the surface of the land.

27·5° may not sound as if it were much of a slope, but in fact it is a considerable one; it is just under 1-in-2. Porlock Hill, Devon, at one time the motorist's terror, 'like the side of a house', is 1-in-3 and 1-in-4, equivalent to angles of about 18·5° and 14° respectively.

STEEP RHINELAND VINEYARDS

There are south-facing, terraced vineyards on the Rhine with 1-in-2 slopes and even steeper. They are impossible to work with ordinary machines and workers in them have to be roped to supports to prevent falls. They are thus much more expensive to cultivate and soil erosion is a problem too, as the earth tends to wash from the top to the bottom. At times these washings are carted back to the top again. The insolation is considerably increased together with the quality of the wine produced and possibly the extra cost involved is paid for by the improved quality.

It should be noted that the rows of vines, or the individual 'goblet' plants on these steep Rhineland slopes are not growing at right-angles to the slope itself, but vertically (at right-angles to level ground or the water in the river), consequently the sun does not beat down directly on top of them, but shines down slightly on the southern side. Nevertheless the slope greatly improves insolation, heating and thus photosynthesis; the land itself is at right-angles to the sun at noon on Midsummer Day (and nearly at right-angles to it in the weeks before and after that day) so the soil is much heated by the sun's impact, leading to quality in the grapes. The micro-climate has been greatly improved.

Northern summers

It is common knowledge that in northern latitudes the summer days are longer than those in southern parts and that consequently the sun rises to the north of east and sets to the north of west, thus shining for a while on the northern side of vine rows with an east-west orientation. The amount of insolation the rows get depends on latitude, whether the ground is sloping or level, the height of the vines and the spacing and direction of the rows. Obviously closely spaced rows shade one another.

SLOPE SHADOW

Level ground is illuminated as soon as the sun rises, but in a vineyard with rows running north–south (the usual orientation, though not necessarily the best) only the external row would be lit at sunrise and only then if no hedge or windbreak cast a shadow. Consideration of the amount of insolation on level ground is dealt with below (page 60).

On sloping ground things are different, because of the phenomenon of 'slope shadow'. Obviously the orientation of a slope makes a big difference—a western-facing incline is not going to be illuminated by a rising sun, whereas level ground would be lit by it: but some shading effect is also to be found on south-facing land. If you watch the sun rise or set from a high point among undulating hills you can see the phenomenon of slope shadow occurring before your very eyes. As an example, the very steep Uckfield southern slope (27·5°) postulated in Fig. 2, giving directly impinging sunlight at the zenith (around noon) on Midsummer Day, would not be lit by the rising sun on that day until quite a while after sunrise, in fact not until about 7:30 a.m., the sun having risen at about 3:45 a.m. (Greenwich Mean Time, not Summer Time). However, if the slope were but 10°, a much more likely figure, our Uckfield vineyard would be lit by it just after 5 a.m. Slope shadow is one of the *disadvantages* of sloping land.

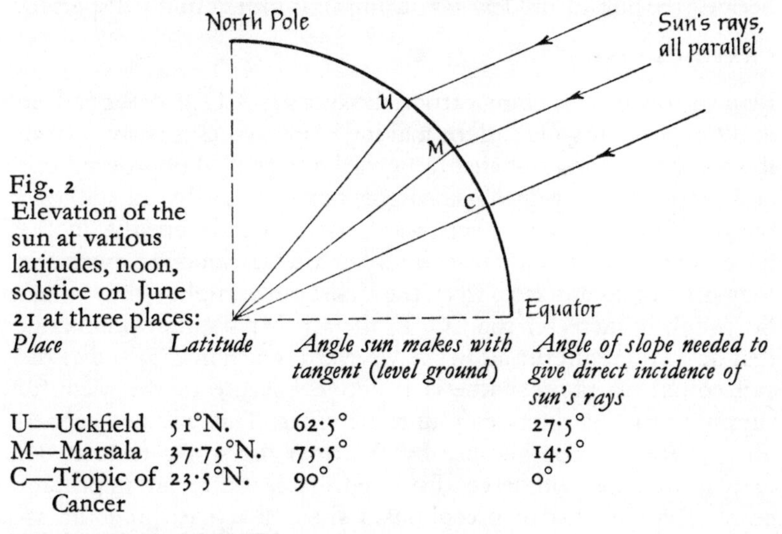

Fig. 2
Elevation of the sun at various latitudes, noon, solstice on June 21 at three places:

Place	*Latitude*	*Angle sun makes with tangent (level ground)*	*Angle of slope needed to give direct incidence of sun's rays*
U—Uckfield	51°N.	62·5°	27·5°
M—Marsala	37·75°N.	75·5°	14·5°
C—Tropic of Cancer	23·5°N.	90°	0°

MR. W. B. N. POULTER

Mr. W. B. N. Poulter, who has a fine vineyard in the Isle of

Wight, has studied insolation on slopes and level ground and has kindly allowed me to print his conclusions, some of them surprising but quite true. They will be found in Appendix IV (page 175). For me two of the remarkable points are that (i) a due westerly or due easterly slope is inferior to a level site in insolation and (ii) in midsummer a very steep slope (40°) gets less sun than a fairly steep slope (30°), showing the importance of slope shadow.

In practical terms the moral seems to be that if you cannot plant your vineyard on a southerly, or near southerly slope, then level ground is best. A 40° slope is, of course, extremely steep—very nearly equal to a hill of 1-in-1.

Practical considerations

In addition to the theoretical aspects of sloping and level sites there are a number of practical ones as well; for instance, morning mists and shelter. An easterly slope in an area subject to morning mist will get less sun than a westerly one because the early sun will be cut off by the mist; mist will not form on a westerly slope until after sunset. On the other hand heavy dews will be dispersed more quickly on an eastern slope than on a western one and thus the eastern aspect will be less liable to mildew attack than the western because the humid, mildew-favouring air is more rapidly dispersed.

CARBON DIOXIDE

Plants grow by absorbing carbon dioxide gas (CO_2): all the carbon in the sugars for wine and for making plant tissues is derived from the air. This process of absorbing carbon is called photosynthesis and can only occur during daylight, not necessarily full sunlight; but it stops at night when plants continue to breathe in the normal way. This is the reason nurses always remove flowers (or they used to, at any rate) from the sickroom at night—the vegetation slightly increases the CO_2 in the air. Mr. Poulter points out that on a still night the air in the vineyard is rich in CO_2 and as the sun comes up photosynthesis is active because of the plentiful supply of carbon dioxide around the vines. This will not happen on a westerly slope because by the time the sun gets to it the carbon dioxide will have dispersed. It is quite an important point. Experiments in greenhouses show that enriching the atmosphere with CO_2 improves growth, thus vindicating the lazy gardener of Victorian times who said it was not necessary for him to repair the hot-house's leaky chimney, because 'the smoke really feeds them early seedlings'.

Hedges, trees and hills, naturally, can also shade the vines from the early and late sun, and so can the vines themselves, which brings us to the subject of the direction of the rows, and the distances between rows.

Direction of rows

NORTH AND SOUTH

Let us take a vineyard on level ground at, say, Bletchley, Bucks. (52° N), with the rows running north and south, planted 5 ft apart and rising to a height of 5 ft, and examine the insolation for 21st June. There on that day the sun rises at 3:39 a.m. G.M.T. and sets at 8:24 p.m., a period of 16 hours and 45 minutes (16·75 hours). While the sun is easterly and westerly and below an elevation of 45° one row will be shading another, only just doing so when it is at 45° but to a greater extent when it is below that height. When the sun is higher than 45° all the vertical sides of the rows are illuminated.

The sun rises to a zenith elevation of: $90 - 52 + 23{\cdot}5 = 61{\cdot}5°$.* Consequently it will be 45° high in

$$\frac{8{\cdot}38 \times 45}{61{\cdot}5}$$

$= 6{\cdot}13$ hours from sunrise, that is at 9:47 a.m.

Of course the rows again shade one another in the afternoon, for the same period of 6·13 hours, that is from 2:15 p.m. ($8{\cdot}38 - 6{\cdot}13 = 2{\cdot}25$ = 2:15 p.m.). These two times (9:47 a.m. – 2:15 p.m.) bracket the times when the whole vertical face of the rows (the whole height) is receiving sunlight, except for a brief period around noon when the sun shines right down the row in moving from shining on the east sides to shining on the west sides. Of course, the top portions of the vines are illuminated before and after these times.

Under the conditions postulated above the upper halves of the vertical faces of the rows will be lit from the point where the sun has risen to an elevation of 27°, that is from 7:20 a.m. onwards. The shadow gradually drops until by 9:47 a.m. the whole vertical face is lit.

* 90° North Pole, 52° Bletchley, 23·5° Tropic of Cancer.

In the afternoon the reverse process occurs. The shadow of one row on the next starts creeping up the vertical face from 2:16 p.m. onwards and is half way up at 4:43 p.m.

At this latitude (52° N) then the north–south rows are fully illuminated (first on the east side and then on the west side) for some four and a half hours and there is an additional period of just under five hours when at least half (the top halves) of each row is lit. It should be remembered though that, except in the Lenz Moser and double curtain systems, most of the fruit is on the lower part of the vines. However, the lit leaves on the top halves will be working actively, making food for the fruit, and some sunlight will be reflected downwards energizing lower leaves and warming the young fruit and the soil.

The above remarks apply to rows within the vineyard. The outermost rows and vines are different. The easterly row gets its whole face lit all the morning and the westerly row all the afternoon, unless shaded by hedges or windbreaks.

As noted above, a certain amount of shadow is cast by the midday sun along the north–south rows. The southernmost vine of a row gets full sun and shades the next but southernmost vine in the row, each vine shading its neighbour all the way up to the northernmost plant. Also the topmost leaves shade the lower leaves. However, leaves do tend to arrange themselves in a pattern which permits the maximum penetration of light, and, it must be remembered, full sunlight is not required for photosynthesis. The sunlight also heats the plants and soil, an important factor for growth and ripening, as reflection from the soil tends to warm the grapes. The figures above are for the longest day; the periods will be less for other days.

Some recent experiments in Georgia (Russia) showed that under their conditions photosynthesis was greatest in the morning and that at midday it was at its maximum in 35 per cent of full sunlight: the figure for the afternoon was 75 per cent of full light. The figures seem very low, but then it was for a lower latitude than ours and under a fiercer sun.

EAST–WEST ROWS, 5 FT HIGH, 5 FT APART AT 52° N (BLETCHLEY, BUCKS.), 21ST JUNE

Some consideration must now be given to rows running east and west, again taking them to be 5 ft high and 5 ft apart. The sun, rising at 3:39 a.m., will illuminate the northernmost row of the vineyard but none of the others, as each row will be shaded by its northern neighbour. By the time the sun has risen to 27°, which will be at 7:20 a.m., it will be illuminating the top halves of the southern faces of the rows from the south-easterly quadrant.

The whole southern sides of the rows will be lit when the sun is at 45° above the horizon, which, as before, will be from 9:47 a.m. to 2:15 p.m. After that shadow starts to creep up the rows and the lower halves of the rows (except the most westerly one) will be in shadow from 2:15 p.m. till the next day. There are thus four and a half hours per day when the rows are fully lit (for the days around midsummer) all from the southern side. The northern sides of the rows will get very little illumination. By about 6 a.m., when the sun will be due east and changing from lighting the northern side to doing so on the southern one, the sun will be 17° over the horizon and will have been lighting some of the upper portions of the northern sides of the rows for a short while, the greatest amount being just over 2 feet.

Latitude naturally makes a difference to these figures, but if the latitudes are near to the Bletchley one the difference is not great. At lower latitudes, though the zenith is higher, sunrise is later and sunset is earlier. For instance, at the Lizard, south Cornwall, 50° N, the sun's zenith will be 2 degrees higher, but sunrise, correspondingly, is a little later and sunset a little earlier. Nevertheless the Lizard vineyard would have the sun at 45° about 10 minutes earlier, with another 10-minute advantage towards the end of the day.

These times are differences in terms of the Greenwich meridian. As the Lizard is 5° 13′ west of Greenwich the real time there is 23 minutes later than Greenwich, although Greenwich time is actually used in everyday life. (Come the Republic of West Penwith, of course, things might be different!) Without the latitude effect, at the Lizard the sun rises and sets 23 minutes later than at the Greenwich longitude (the meridian) and the latitude takes off another 10 minutes at each end.

If there are two or more pickings at harvest, or if *auslese* wines are to be made, the east-west direction could be advantageous because the south-side grapes would be of high quality and once they had been removed the sun would start passing through the row, warming up the north-side ones. Moreover, the fruit will tend to form on the south side rather than on the north so the south side will carry most of it. John Rose favoured the east-west direction for this reason and also because it gave more protection from the south-west winds. It may sound like heresy to the north-south vignerons but there is much to be said for an east-west direction of the rows.

Windbreaks or sheltering hills may be important to a vineyard but their advantages must be set against the fact that they may cut off considerable amounts of sunlight, particularly from the rising and setting orb.

Wide and close planting

How far apart should the rows be? It is said that closely planted vineyards give good quality because of the root competition and it is a fact that many poor lands give fine wine. Edward Hyams has expressed it very neatly. He said it was as if there were only a certain amount of quality in any given bit of land (say an acre) and, according to how you treated that land, you could spread that quality over 1,000, 2,000 or 10,000 bottles. This is why the big French growths, generally on poor thin soils, have the quantity they may produce limited by law to certain fixed, and usually small, amounts. Thus close planting with its resulting root competition tends to reduce the volume of wine produced per acre and to improve quality.

Closely planted vineyards are more difficult to work and harvest: special machinery must be used for cultivation, or even hand tools only. Wider planting allows ordinary farm machinery to be employed and permits a better penetration of sunlight to the vines and soil, but the production per acre may be reduced without any corresponding increase in quality from root competition. On the other hand there is a potentially greater quality from a better penetration of the sun. The Lenz Moser system of growing is designed to make the whole working of the vineyard easier, thus making it more attractive to workers in that field. Labour is the expensive commodity today.

The Lenz Moser system

In Dr. Lenz Moser's method the rows are 10 ft 6 in. to 11 ft 6 in. apart and the spacing in the rows is from 3 ft 4 in. to 5 ft apart, from 770 to 1,050 plants per acre being needed, instead of the

1,742 vines used in a more conventional 5 × 5 ft system—a considerable economy when buying plants.[68] Posts 8 ft long are driven about 18 inches into the soil and the twin top wires are at 6 ft 5 in. above the soil. Another row of double wires is below this, at 5 ft 5 in., and at 4 ft 5 in. above the soil there is a single wire. The vine is trained to produce a shoot reaching to the lower wire and it then branches along that wire in both directions. All the growth and fruiting is along this wire; some of it hangs down, forming a 'curtain', and some of it goes up, being trained between the two upper twin wires. Since all the pruning and most of the fruit itself are at breast height, these operations are comparatively easy. Two possible disadvantages are that it may take an extra year to establish the fruiting wood system and that the fruit is well away from the soil and the heat reflected from it. But to some extent the wider spacing may compensate for this by allowing better sun penetration. Dr. Lenz Moser's objective all the time he was experimenting with different methods was to make the working of the vineyard easy and thus cheaper. He maintains that with the conventional system from 1,010 to 1,030 hours of work per acre per year are needed, whereas with his method only 170–240 hours per year will be called for. The method, he maintained, was very suitable for small, family vineyards; for instance a family with two working adults could only cultivate 1 hectare (2·5 acres) of vines under the old system but, provided with a suitable tractor and sprayer, the Lenz Moser system would enable them to handle 5–7 hectares (12–17 acres), making the small vineyard a much more viable proposition. The author does point out, though, that extra labour will be needed for the harvest.

Dr. Lenz Moser also quotes some interesting figures on machine travel per hectare for the various distances between rows; that is the length of run of the tractor when ploughing, cultivating or spraying. The figures, converted to yards run per acre, are:

Rows	3 ft 4 in. apart	:	run per acre	3,717 yds
	4 ft 11 in. apart	:	run per acre	2,965 yds
	11 ft 6 in. apart	:	run per acre	1,283 yds

that is, over 2 miles for the closest planting and under three-quarters of a mile for the widest, a considerable saving for the latter distance. Expressed in this way the system sounds most attractive, but one needs to ask whether with this wide spacing the yields per acre are any lower. In the English edition of his pamphlet Dr. Lenz Moser does not have much to say about yields but does mention that his system will allow mechanical harvesting, in which case varieties giving long-stemmed berries yielding 10,000 kg per hectare should be used. This is equal to over 4 tons per acre, a quite respectable crop. He also has a general remark that if the right varieties are used the crop is much higher with his method than with close planting. Even were the yield per acre reduced the savings from using the system would more than compensate for it.

Double curtain system

The double curtain method is an adaptation of the Lenz Moser scheme and a splendid vineyard planted this way is that of Mr. B. H. Theobald at Westbury Farm near Reading, Berkshire. A more elaborate wirework is needed but it uses yet fewer plants per acre than the Lenz Moser system. Stout wooden T-shaped supports about 5 ft high carry wires at each end of the cross piece of the T. The rows of T-pieces, running north and south, are about 12 ft apart. In the rows vines are planted at 8-ft distances, 453 plants per acre being needed. Canes are inserted beside each plant, one leading to one side wire (say the eastern wire of the pair) and the next cane to the other side wire (the western one) and so on till the end of the row. Wood from each vine is trained along its cane and then, turning at right-angles, along the wire in each direction (north and south) for 8 feet. From these wires the growth and fruit tend to hang down, forming two curtains. It is said to be very successful, and certainly Mr. Theobald gets good crops of excellent wine. Mr. A. H. Holmes has also planted a vineyard on this system at Wedmore, Somerset.

Yields from different systems

Comparative yields from different training systems are reported in the horticultural literature from time to time. For instance, in the Crimea a hectare of fan-trained vines was turned over to the Lenz Moser system and over the next seven years it not only produced 37 per cent more fruit but the grapes also had a higher sugar content. At Plovdiv, Bulgaria, a high-stem wide-planted vineyard ('Cabernet-Sauvignon' on Kober 5BB) had a number of plots under different pruning systems. The following figures were obtained:

Pruning system	*Yield kg per vine*	*Per cent Sugar*	*Sugar per vine (kg)*
Modified Lenz Moser	2·040	22·7	0·46
Double curtain	2·950	23·2	0·68
Umbrella	2·570	23·1	0·59
Modified Kniffen	2·676	22·6	0·60
Two-tiered single cordon	2·933	22·5	0·66
Two-tiered double cordon	1·607	21·8	0·35

Under the Bulgarian conditions the double-curtain and two-tiered single cordon systems did very well.[90]

Soil colour and cover

Soil cover is another factor: root activity is greater in warm soil than in cool and the sun illuminates and warms not only the foliage but also the soil. Dark soils absorb heat more rapidly than light ones and vine growers in the Moselle often cover the soil round vines with slate so as to increase soil temperature. What may be forgotten, though, is that dark surfaces also radiate heat more rapidly than light ones, and consequently cool more rapidly at night. However, since in summer in northern latitudes the day-light hours are longer than the night hours the dark soils probably show a net heat gain over the light ones. In fact Mr. Poulter quotes some Bedfordshire figures on this point.[89] In an experiment on neighbouring plots of soil at midsummer, temperatures at 10 cm depth were: medium loam 18°C, soil mixed with soot 22°C and soil mixed with chalk 15·5°C. Another overlooked point is that if the heat is being absorbed by the dark surface it is not being re-radiated to warm the foliage and fruit. Mr. Poulter also found that cover crops of grass, clover, lucerne or even weeds affected absorption of heat by the soil; on the whole such cover, if not excessive, simply delayed the rise in soil temperature by about four days. Once the soil was warmed the cover crop protected it from changes during short cool spells. Cover crops of this nature have the advantage of being more easily worked: all that is needed is a few mowings a year instead of ploughing, cultivating or applications of weedkiller.

Of course, there is a band of land directly beneath the vines, a foot or 18 inches wide, which must be weeded, a subject dealt with in Chapter 7.

M. B. Juillard recently put forward the view that a cover of vegetation between the rows (he was referring to weed growth) altered the micro-climate by its transpiration, its own radiation and its insulating properties between soil and air, making such sward vineyards more liable to damage by spring frosts, especially those on flat land and in valley bottoms.[53] However, M. Juillard also pointed out that working the land in spring to kill this cover also momentarily increased the risk of frost damage, thus leaving the vigneron on the horns of a dilemma as to the action he should take.

Varieties

Vine varieties fall into two classes: (i) Pure *Vitis vinifera* cultivars, obtained by crossing different varieties of this species or by isolating and multiplying a bud sport; and (ii) hybrids obtained by crossing *vinifera* with one or more American species.

In France the crosses between pure *vinifera* varieties are known as *métis*, as distinct from the inter-specific crosses called *hybrides*. In the vine-variety sense there is no real translation to English of the word *métis*, unless it be the rather clumsy '*infra-vinifera* crosses': the common meaning of the word is 'mongrel' or 'cross'.[20]

The new hybrids are not popular on the continent, at any rate among governments and the European Economic Commission, for two main reasons. The hybrids are usually more productive than the *métis* vines and hence threaten to lower prices for the established vignerons; secondly, being resistant to some or all of the pests, they threaten the pesticide trade, and, as far as phylloxera goes, specialist nurserymen selling grafted vines. If grafting against the phylloxera is not needed, because the hybrid itself is resistant, at any rate to the root form, then the making of new plants by means of layering or the taking of cuttings is easy: specialist vine nurserymen would be out of business. Resistance to the mildews means no spraying, or very little of it—a disaster for the chemical trade. The hybrids are also known as 'direct producer vines' to distinguish them from hybrid rootstocks, made by crossing American species. In addition there is the question of whether the hybrids introduce traces of the terrible foxy flavour into the wine, characteristic of wines from species of American grapes—a much debated point. Certainly some do; 'Brant' is an example, but the *labrusca* flavour in it is so slight that it is not disagreeable. On the other hand 'Seyval Blanc' (Seyve-Villard 5,276), now a popular variety in England, shows no trace of its American descent in the wine. In blind tastings in France wines from hybrids have gained top marks. An association exists there to study the matter and to look after the interests of the growers of the 'direct producer' wines. It is FENAVINO (*Fédération Française de la Viticulture Nouvelle*), Poitiers, France. Needless to say the opponents of the hybrid vines make much of the threat of 'fox' in the wine.

The valuable work done at the Oxted Viticultural Research Station has already been mentioned; as early as 1961 Mr. Barrington Brock had already picked out as promising two of today's popular varieties, namely the 'Riesling/Sylvaner' (now more commonly known as 'Mueller-Thurgau') and Seyve-Villard 5,276.

Britain now comes under the Common Market wine regulations. The European Economic Community (EEC) has published a list of varieties, recommended and authorized, for cultivation in the United Kingdom.[115]

Varieties recommended: 'Auxerrois', 'Mueller-Thurgau', 'Wrotham Pinot' ('Pinot Meunier' or 'Malinger').*

Authorized: 'Bacchus', 'Chardonnay', 'Ehrenfelser', 'Faber', 'Huxelrebe', 'Kanzler', 'Kerner', 'Madeleine Angevine', 'Madeleine Royale', 'Madeleine Sylvaner', 'Mariensteiner', 'Ortega', 'Perle', 'Pinot Noir' ('Spätburgunder'), 'Ruländer' ('Pinot Gris'), 'Seyval Blanc' (Seyve-Villard 5,276), 'Seigerrebe'.

Mr. J. L. Ward, Chairman of the English Vineyards Association, writes: 'Although it may be hazardous to suggest other varieties, which have not been fully tested in the prevailing climatic conditions of the United Kingdom, the following should perform equally well in the Southern part of the country:

'*White grapes*

Early Précoce de Malingre, Perle de Czaba.

Mid-season Albalonga, Chasselas, Gutenborner, Rabaner, Regner, Reichensteiner, Septimer, Würzer.

Late Schönburger.

* Badell (1952) gives 'Goujan', 'Auvernat Gris' and 'Gris Meunier' as synonyms.[3]

'*Black grapes*

Mid-season Blue Portuguese, Zweigeltrebe.'

Vines not on the recommended or authorized list may still be planted, but for the time being will be regarded as experimental. As the quantity of wine produced from English vineyards is so minute compared with the total European production, the EEC would be unlikely to order the rooting up of a vineyard planted with an unauthorized cultivar, but it might ban further plantings of it.

The wine grower and the would-be vigneron thus have a considerable choice of varieties before them, which is a very good thing, because it is *variety* that is wanted in a nascent industry such as this. The comparison of single-grape wines with wines from blends of grapes, of vintages from one area with the same grape in another and of one year with another, is an unfailing subject of interest to wine producers and drinkers.

The two most planted varieties in England today are 'Mueller-Thurgau' and 'Seyval Blanc'. The first part of the first name is spelt in various ways—Muller and Müller being used at times and the hyphen inserted or dropped. Probably *Mueller* is the best spelling for England. It approximates to the German pronunciation and avoids the difficulty of getting an umlaut inserted over the u. The variety is also known as the Riesling-Sylvaner cross but it must be remembered that there are also many other crosses between these two cultivars.

'Huxelrebe', now cropping in some areas and earlier than 'Mueller-Thurgau', gives a pleasant, flowery wine, somewhat reminiscent of elder flowers. The remarkable summer of 1975 will enable a sounder judgment of the possibilities of the new varieties to be made when the year's wines are ready to drink. There is a suggestion around that the EEC will ban the planting of 'Seyval Blanc' on the grounds that it is a hybrid. It must be remembered that the Commission is haunted by the spectre of the 'wine lake', of having to buy up vast wine surpluses and distil off the spirit in them for industrial uses. Perhaps one day wine will be a considerable substitute for petrol and the hybrids welcomed!

On the continent of Europe at present hybrids are banned from vineyards selling wines under the *Appellation contrôlée* classification and no doubt a similar prohibition will be imposed on English vineyards if and when a similar quality control is established here.

The advantages of some of the hybrids are considerable. They may not need to be grafted, though this is not necessarily an advantage in Britain and is discussed below. Also, being resistant to the mildews, they may not need spraying, or may need only a few treatments—a considerable saving in cost and labour.

Miss Gillian Pearkes, who has an experimental vineyard near Dulverton, Somerset, and might be described as the modern Barrington Brock, sees hybrids as only suitable for amateurs making and drinking their own wines. Professional producers, she thinks, should adhere to the noble *viniferas*, but the temptation to plant trouble-free hybrids is very great. In her excellent book she lists some thirty-eight white and five black *viniferas* and two white and four black hybrids.[84] I am indebted to that work for some of the following notes.

My personal selection of varieties for a Hertfordshire vineyard would be for the main part to be half and half Mueller-Thurgau and Seyval Blanc, and then a few rows or half-rows of the other varieties mentioned below. They are arranged in alphabetical order. Larger plantings could then be made of any showing marked advantages.

White Grapes

'Albalonga'. A very vigorous grower, high sugar content, slow maturing wine.

'Huxelrebe'. More vigorous than 'Mueller-Thurgau', does not like chalk, susceptible to both mildews and *Botrytis*. Sugar and acid higher than 'Mueller-Thurgau'.

'Madeleine Angevine 7972'. Vigorous, crops more heavily than 'Mueller-Thurgau' and ripens earlier. Prone to mildew. Probably suitable for poorer sites and areas. Muscat-type wine.

'Madeleine Sylvaner 28/51'. Ripens early, hence suitable for the less favoured sites: smaller crop than 'Mueller-Thurgau'.

'Mariensteiner'. Vigorous, pollinates well. Resists spring and autumn frosts. High sugar and acid.

'Mueller-Thurgau'. (Raised by Herr Müller at Thurgau, Switzerland.) Susceptible to *Botrytis* and powdery mildew. Makes an early maturing, comparatively low acidity wine. Mid-October ripening.

'Perle'. (Würtzburger Perle: do not confuse with 'Perle de Czaba'.)

Highly resistant to winter and spring frosts: has stood up to −6°C. in May. Ripens a little after 'Mueller-Thurgau'. Susceptible to *Botrytis*. Modest sugar and low acidity. This could well be a variety worth having in a modest amount, as a standby in a year of spring frost. At least some wine could be made that year.
'Regner'. Needs a well drained, but not chalky, soil. Resists powdery mildew but is susceptible to the downy mildew.
'Seigerrebe'. Avoid chalk. Early cropping (hence liable to attack by wasps). High sugar and low acidity. Strong flavour and considerable bouquet to the wine.

Black Grapes

I would not take the black grapes very seriously, but would plant a few just to see how they did.
'Siebel 13053'. A hybrid, resistant to the mildews. Medium to light cropper. It is a *teinturier*, i.e. the flesh is coloured so it cannot be used for white wine, only for *rosé* or reds.
Vitis purpurea. A red-fleshed grape, useful for adding colour to red wines.
'Wrotham Pinot'. Medium to light cropper, a clone of 'Pinot Noir'.

A number of Russian grape cultivars have created interest, such as 'Gagarin Blue', 'Kuibishevsky' and 'Tereshkova', probably on the general idea that Russia is a cold country and that their varieties should well suit us. I think this view is mistaken. The Russian viticultural area is on the shores of the Black Sea—the Crimea, for instance, and Georgia. The climate is not the least like ours. The region is much farther south; Sevastopol is 44° 37′ North. Secondly, summer insolation is stronger and the summer temperature is higher. Thirdly, the winters are cold enough, much colder than ours, and the Russian vines are bred to resist winter cold, a factor we do not need to consider with our comparatively mild winter.

It must be kept in mind that most of the new varieties are protected under the Plant Varieties Act and must not be reproduced unless permission has been obtained from the owner of the rights in those cultivars—usually the original breeders. Such a person, usually a nurseryman as well as a vigneron, has put much effort into creating new varieties and it is only just that he should reap some benefit from the propagation of a promising new vine by collecting a royalty, or by keeping the new cultivar to himself. At the same time it must be remembered that such vine nurserymen have an interest in seeing the old varieties replaced by new ones. The following are among the protected varieties:

'Albalonga'	'Huxelrebe'	'Ramdas'
'Aris'	'Kanzler'	'Reichensteiner'
'Augusta Louise'	'Kerner'	'Regner'
'Bacchus 113'	'Kolor'	'Rieslander'
'Domina'	'Mariensteiner'	'Septimer'
'Ehrenfelser'	'Optima 115'	'Wurser'
'Faber'	'Ortega'	No. 3001-9-129
'Forta 100'	'Perle'	No. 3146 GM.
'Freisamer'	'Rabaner'	Trollinger × Riesling

Stocks

Should the modern vineyard in Britain be set out with grafted or own-roots plants? It is a considerable question. The object of grafting is that the plants are then immune to phylloxera—the American insect which nearly wiped out *vinifera* throughout the world a century ago. I have told that story in another book.[80]

However, phylloxera does not exist in Britain, and here grafting is not essential to vine culture as it is in mainland European growing. It is merely a precaution against the introduction of the pest. Grafted plants tend to produce slightly bigger crops and to have shorter lives than own-roots *viniferas* and, of course, the former are more expensive.

The process of grafting is quite elaborate and is done in the reverse manner to what one might expect—the join between stock and scion is made *before* roots are induced on the stock. Usually the whip and tongue graft (or *greffe anglaise*) is used. The work is usually done in March. Short lengths (25–30 cm—10–12 in.) of suitable American wood are selected and the two cuts are made just below a bud. The scion consists of one bud in wood of the same diameter and the corresponding two cuts are made again just below the bud. If made by a machine (as is now more usual) or by a skilled grafter, the stock and scion can now be fitted neatly together. At one time the stock and scion were tied round with raffia and then painted with warm grafting wax, but this is no longer found to be necessary. The joined grafts are now 'stratified', that is, packed into wooden boxes with a special 'jointing medium', usually damp sawdust with a little charcoal and a plant growth hormone. The filled boxes are put into a warm room (about 30°C). On the seventh and twelfth days the boxes have to be plunged into warm water because they tend to dry out. Usually the boxes are handled so that only the stocks are wetted and not the scions. After about fifteen days at 30°C the boxes are taken out to a cooler room and after twenty-one to thirty days the process ends. The grafts are now examined. Most of them will have joined, a jelly-like growth forming round the actual graft. Any shoot growth from the stock is rubbed off and the joined wood is then planted out in rows in shallow trenches in open ground to form roots. In July they are examined; if any roots have developed from the scion these must be rubbed off because, obviously, they could be a focus of phylloxera infection.

The multiplication of own-roots plants is far simpler, layering and cuttings being the two methods. A growing shoot buried in the soil up to, say, July, will strike roots from the nodes and send up an aerial shoot from each one. Cuttings of three or four buds should be taken from ripened wood in early autumn (before all the plant food in the stems has drawn back into the rootstocks), treated with a rooting hormone and planted into shallow trenches. Most of them wll strike roots in the following spring.

The comparative ease and difficulty of the two methods are reflected in the prices of grafted and own-roots vines; the former sell from £250 to £350 per thousand and rooted cuttings cost about £200 per thousand.

A few grafted plants, with apparently sound joins, often die for no apparent reason and examination may show that the union has failed, wind or vibration putting too great a strain on the joint.

The grower thus has to make a choice between using comparatively expensive grafted vines and being safe from phylloxera or using own-roots plants, which could slowly be wiped out should the pest be introduced. The subject is discussed more fully in Chapter 5.

The actual stocks most commonly used in Britain are Kober 5BB and SO4.

Spring frosts

The usual precautions against spring frosts must be borne in mind when selecting a site for a vineyard. Cold air flows downhill and can be held on a slope by tall, thick hedges, woods, even buildings or a wall. Such cold, stagnant air can cause great damage to the young spring growth and decimate the crop. A site should be selected which provides drainage for this cold night air. Obviously valley bottoms should be avoided.

Direct methods against spring frosts consist in heating the air and, strangely enough, spraying with cold water. Matting can also be spread over the rows in the evening and removed in the morning, but the process is extremely laborious. It used to be widely used in the Champagne. Yet a fourth system is to grow a frost-resistant variety, such as 'Perle'.

The selection of a safe site is obviously the cheapest and most satisfactory method of frost control. Smoke clouds are now not thought to be of much use as regards the smoke itself. Any protection given by them is due to the heat generated rather than the smoke particles, and if the smoked vineyard is anywhere near a town or village many protests will be received. Orchard heaters usually use fuel oil, which has now become too expensive. Spraying with water is the easiest method, but can really only be done satisfactorily if the vineyard is fitted with an overhead sprinkler irrigation system—an expensive outlay. When the alarm thermometer rings the 'danger' bell the water is turned on. The vines and young growth will soon be covered with ice. The latent heat of the freezing process then protects the vines from falling to the danger point of about −2°C. The objection is that the system is expensive to install, especially as in most seasons the vines will not need irrigating. The vine is deep rooted and seeks its own water.

Spring frosts can be very troublesome to all fruit growers. In Hertfordshire, on the night of 30 April–1 May 1975, the ground temperature fell to −5 °C, doing extensive damage to vines. Frost-damaged vines should be pruned back hard to induce the basal buds around the shoots to break.

Hail

Hail when the grapes are ripening can cause much damage. Firing rockets to break up hail clouds is expensive, spectacular and great fun, but it is not known really if it is effective. The nets put on vines against birds break the fall of much hail and give considerable protection to the fruit.

Pollution

A vineyard should not be alongside a main road for the fruit can pick up a considerable quantity of poisonous lead from motor exhaust gases.

Some tests made at Trieste, Italy, showed that the lead content of ripe grapes next to a main road was from 4 to 6 parts per million parts (ppm) of the dried grapes.[61] Since grapes are about 78 per cent water that is equivalent to about 1·1 ppm of fresh grapes, not a very high contamination, but undesirable. However, at 50 metres distance from the road the lead content of the fruit was negligible and at 100 and 150 metres it was nil.

Professor G. W. A. Fowler made some similar tests on blackberries and elderberries in England.[35] Blackberries did not pick up much lead (0·14 to 0·85 ppm) but some elderberries within 8 feet of the A4 picked up 6·77 ppm of lead. Other elderberries, having 4·63 ppm of lead, and made into wine, using them at the rate of 400 grammes of fruit per litre, gave a wine with 1·23 ppm of lead. This was from unwashed fruit: when the elderberries were washed the lead content of the fruit was reduced to 1·45 ppm and of the wine to 0·4 ppm. The higher level is not very bad, but even so is undesirable. A man drinking a bottle a day of the high-lead elderberry wine would take in over 800 μg per day—twice the tolerable lead level, which would be serious; but if he were drinking a bottle a day he would have made a considerable quantity of wine and not all his fruit would have been taken from the immediate roadside: the lead contamination would in practice be less.

Similarly with the Trieste grapes, 1·1 ppm lead in the grapes could easily give the same figure in the wine. The consumer could thus get twice the tolerable dose by taking a couple of glasses of that wine a day. But again it is only the first row or so of grapes that would be contaminated. It was obviously a big vineyard, stretching at least 150 metres from the road, so the lead-laden grapes would be well diluted.

Nevertheless, it is as well to avoid being too close to a road; in addition to lead accumulation there is the possibility of undesirable taint from exhaust fumes and scorching from ozone emission.

Lead poisoning has worried the French authorities for some time. They even have a special word for it, *saturnisme*. Wine was often the source of the poison. M. E.-J. Armand Gautier, in the 1880s, pointed out that some café-keepers and wine merchants would drop a piece of metallic lead, or some litharge (protoxide of lead), into each barrel of wine to act as a preservative, or to correct acetification, converting the liquid into a 'veritable poison'.[37] Lead could also be picked up from the solder joining copper pipes carrying wine.

Pruning

The object of pruning is to secure good crops of well-ripened fruit, consequently some consideration needs to be given to the botany and physiology of the vine.

Vitis vinifera was originally a dioecious plant, that is a species having separate male and female flowers on different plants. The female flowers had short deformed stamens, producing no viable pollen, and the male flowers had inoperative pistils, generally incapable of development. In the wild, wind or insects carried the pollen from the male flowers on the male plants to the female flowers on the female plants, where grapes then appeared. But a few hermaphrodite flowers occasionally occurred, having both viable stamens and pollen in the same blossom. When man interfered and upset 'the balance of nature' by inventing viticulture, the hermaphrodite flowers were automatically selected.

Having accidentally discovered the delights of wine, the early would-be vigneron set out a vineyard. About half the plants would have been male and the other half female, but mixed in would be a few hermaphrodites. Our grower would soon have found that about half his vines (the males) gave no fruit, so he would eliminate these, whereupon most of the rest of the plantation—the female plants—would give no or very little fruit because of the lack of pollination. It can be a great mistake to cut down the barren fig-tree and cast it into the fire. Even in the vegetable world males have a few uses.

Disaster appeared to face our early vigneron. What had he done? Were the gods angry at his not being content with the wild fruits? Was he being punished for greed, for introducing these new-fangled methods, for disturbing the natural order of the good old days? A tricky question. But here and there among the vines was a bush bearing a fine crop (the occasional hermaphrodite plant), surely a sign—a sign of hope—from his own particular god. By means of cuttings and layers he would have propagated new plants from those fruitful vines and have established in due course a range of hermaphrodite varieties. Today all *vinifera* cultivars used in viticulture bear male and female organs in the same flower; they are mostly pollinated by the wind, though Mr. J. Free has pointed out that insect pollination of vines, particularly by bees, may be more important than we think.[36] He quotes some Russian experiments in which bees were trained to visit vines by feeding the insects sugar syrup in which vine flowers had been put. Because of better pollination crops were increased by from 23 per cent to 54 per cent.[6] Some varieties of grape, such as the 'Almería' of Spain, still have to be hand pollinated by touching clusters every day during flowering with a bunch of flowers from a pollinizer vine—a laborious task. Occasionally all male or all female flowers are still found among modern *vinifera* cultivars.

The vine fruits on growths arising from stems of the current year's growth. These blossoms arise from points opposite a leaf. Tendrils may also be produced at those points and in fact the bunch of flowers is really a modified tendril. Tendrils may often be found carrying a few grapes.

Thus to induce flower formation we need to prune in such a way that the spring growth is vigorous and can make and carry enough food to induce at least one, if not two, flower bunches instead of tendrils on most of the shoots. Tendrils are also wanted in the vineyard because they save us a great deal of work: much more hand tying would have to be done if the tendrils did not do so much of it for us.

In the majority of the world's vineyards the Guyot system of pruning is used. It is a wood-renewal method and was not really invented by Dr. Jules Guyot, but bears his name because he is the man who much clarified the method and gave it publicity in the nineteenth century.[43] As far back as 1666, as we have seen, John Rose was advocating a very similar treatment. At a pruning competition, 28th December 1863, at Mirande (south-west France) the jury recommended Dr. Guyot's system.

Guyot pruning is now so widely practised that it hardly needs description, but a brief summary is given below.

In Guyot pruning most of the work is done in winter. In the simple system a healthy, well-ripened shoot (No. 1 shoot) is tied down to the bottom wire and cut to fit the distance between the vine and the next plant in the row (say at 5 feet). All the other shoots are ruthlessly cut right away except one, which is pruned to two or three buds (No. 2 shoot). The brutality often breaks the beginner's heart. In the following spring and summer No. 1 shoot will produce growths from the buds along it (Nos. 1A to, say, 1E) and these will be trained upwards and tucked between the middle and top wires. It is these growths (Nos. 1A to 1E) which produce most of the fruit. The growths from No. 2 shoot (Nos. 2A, 2B and 2C, say) will be trained up the stake and may bear some fruit but their main purpose is to provide the replacement wood for next year. The next winter all the No. 1 shoots are cut away. One No. 2 shoot (say No. 2B) is selected and tied down to the lower wire. One (No. 2A) is cut right away and No. 2C is cut to three buds. Thus No. 2B will be next year's main fruit-producing shoot and No. 2C will make the wood for the following year. The process continues, year by year, *ad infinitum*, or rather for the life of that plant, say fifty years for a grafted plant and eighty for an own-roots one. The life of a vine will vary greatly according to variety, soil and stock. A not negligible factor, particularly in England, is the pressure of the nursery salesman, with novelty, wonder vines, or clones of old favourites.

The Guyot method can be extended. For instance, there is a two-tier Guyot system when the wiring is about 6 feet high and a second tier is established by tying one of the No. 2 shoots to the middle wire. Then there is a double Guyot system, single-tier, and a double Guyot, two-tier. This is just a method of economizing in plants. Suppose the rows run north and south, then in the double Guyot each plant has two lengths of fruiting wood tied down each winter, one running 5 feet north and the other going 5 feet south. It may take longer to establish the vineyard in this way but only half the number of plants are required. In Guyot double two-tier a second tier of shoots, made by vine rods running north and south on the middle wires, is established.

The question now arises as to what time of year the pruning should be done. The vine is a plant which makes no callus over wounds; these are sealed by a gradual drying of the tissues. Consequently, if the pruning is done late, the wounds do not dry and as the sap rises in the early spring the vine 'bleeds'. The 'tears of the vine' were quite famous at one time and were used in medicine.

They consist mostly of water and a few mineral salts serving as plant food. Extensive bleeding from pruning cuts—made in, say, March—can be alarming; soil can often be seen to be quite wet beneath a dripping shoot, but the only loss really is a certain amount of water evaporating into the air. The plant food is returned to the soil and probably encourages surface rooting. On the other hand there is a waste of energy in that the plant pumps up a lot of nutrient solution which just drops back into the soil again. If the sight of bleeding vines makes you unhappy it can be avoided by doing the pruning in December and January. I do not think it good to do it earlier, for it is desirable that as much as possible of the plant food in the shoots to be cut off should be drawn back into the stock of the vine and thus be available for the next year's growth. Guyot himself was of the opinion that the bleeding did not matter. He made some experiments in which some late-pruned shoots were left to bleed while on other plants the wounds were sealed with a red-hot iron. There was no difference in yield from bleeding or sealed plants. Late pruning allowed one to see the state of the buds and select the best shoots for retention. The exception to the above is in taking wood to be used as cuttings. It is best to take this in autumn, but from ripened wood, while there are still some food reserves in the stems.

Every effort should be made to keep the main stock of the vine vertical at least as far as the lowest wire. This can be done by tying it firmly to its supporting cane. A straight stock is needed because it helps to support the weight of fruit and foliage and puts less strain on the wire. If the main stem slants up to the lowest wire it just bends under the weight, giving no help at all to the suspension of the crop. In winter all dead snags and tendrils should be cut away so that they do not become breeding points for fungi, such as *Botrytis*.

Spring pruning consists of rubbing off growth from the main stem and from basal buds which may sprout on the crown of the stem. These are removed because we want all the force of the plant to be exerted on the tied down fruiting shoot (No. 1 shoot in the example given above) and the growth coming from the short stem (No. 2 shoot) set aside for the production of replacement wood. This promotes vigour in the two shoots, leading to the production of flowers. If it is not done, a bush will result and fewer grapes be produced. In varieties which tend to tip-bearing the tied down shoot is arched, encouraging the mid buds to break as well as the tips.

Summer pruning, or '*pinçage*' next occupied Guyot's attention. He advocated the pinching out of the leading shoot at two leaves above the second bunch of grapes or at five or six leaves above the bunch if there were only one bunch. Today Guyot's two usually become three leaves and not more than two bunches of fruit are allowed on any fruiting shoot. In vigorous varieties this shortening will force growth from buds in the axils of the leaves which, Guyot said, should then be pinched out just above the second leaf. Today this is a continuous task throughout the summer. If it is not done the foliage will become too thick, forming an ideal environment for mildew. The tendrils hold in some of the growth but nevertheless much tying of stray shoots will also have to be done.

During the summer, growth over the top wire is usually cut back to allow sun and sprays to reach the grapes. The next question is whether leaves should be cut back in late summer and autumn—August to October—to allow the sun to warm and ripen the fruit.

Phillip Miller (1743) was much against this practice: 'I can't help taking notice of the absurd Practice of those who pull off their Leaves from their *Vines*, which are placed near the Fruit, in order to let in the Rays of the Sun to ripen them; not considering how much they expose their Fruit to the cold Dews, which fall plentifully in *Autumn*, which, being imbibed by the Fruit, do greatly retard them: besides, no fruit will ripen so well when entirely exposed to the Sun, as when they are gently screened with Leaves; and by pulling off these Leaves, which are absolutely necessary to prepare the Juices before they enter the Fruit, the gross Parts of which are perspired away by the Leaves, the Fruit must . . . be deprived of Nourishment . . . and thereby render the Fruit worse than it would otherwise be, were the Leaves permitted to remain upon the Branches . . .'[67] It is interesting to note that Miller realized that the leaves were the factory making the food for the fruit, and he wrote before the American mildews had reached Europe.

It certainly seems logical that the food factory should not be too much curtailed. Obviously it is quite safe to remove old and spotted leaves and very young growth. The large leaves are the ones making most of the nourishment for the fruit.

The matter of the optimum amount of leaf in a vineyard is under study in a number of places. The leaf-surface/fruit ratio is important. Dr. Lenz Moser, in Austria, found that vines grew more quickly, more vigorously and set more fruit the fewer leaves were removed. The best leaf-surface/land-surface ratio was from 2 to 2·5m² of leaf per m² of land.[69]

If a high trellis method is being used, such as the Lenz Moser or double-curtain, Guyot pruning cannot be done and a system of spur pruning is employed. The wood is trained along the high wires as already described. Each spring buds along this wood throw shoots, some of which bear fruit; usually the non-fruiting shoots are cut out. The next winter the shoots that have fruited are cut back to form spurs, from which, in the following spring, shoots will spring. The strongest of these are selected for fruiting and the remaining shoots rubbed out, as are any adventitious buds on the main framework of the wood.

Spur pruning can usually be carried on in this way for a long time (twenty years), but a point may come when many of the spurs die out and growth is concentrated mostly on the buds farthest from the roots (tip-bearing). When this happens the framework must be cut away, which will induce new shoots to arise from the base, and a new framework is then built up from one of these shoots. If grafted vines are being used, be careful not to train up a shoot arising from the stock: one does not want *rotundifolia* or *berlandieri* rogues in one's vineyard.

Vines trained up a house or along walls must also be spur-pruned in winter and the framework renewed from the base if tip-bearing becomes too marked.

Soils

The vine will flourish in almost any soil that is not water-logged or salty and poor soils give the best quality wine, a reason why vineyards are often found on thin, stony lands that would grow little else. For instance the soils of such famous vineyards as those of Château d'Yquem and Château Lafite have from 55 to 70 per cent of stones and pebbles. The stones absorb heat during the day and slowly surrender it at night encouraging extension growth, leaf and fruit formation.

Deep rich loams, advocated for almost every crop under the sun, are not suitable for vines. They give excessive shoot and leaf growth and the wine is coarse and does not keep well.

The nature of the soil influences the kind of wine made. Clays are good for highly coloured, strong red wines, heavy with tannin. Chalky soils produce light wines with good bouquets.

The ideal soil is one having good drainage, no 'pan' (that is, an impermeable layer beneath it), and containing silica, clay, chalk and iron oxide; the last may not be so important for those growing white wines. The older French writers took the view that the red oxide of iron improved the colour of red wines, which sounds like sympathetic magic, but may, of course, be a fact. The vine is naturally a deep-rooting plant and its roots will penetrate far into the subsoil in search of the water and plant food it needs.

Chancrin[20] gives the following figures for some famous vineyard soils:

	Percentage of:			
Vineyard	*Silica*	*Clay*	*Chalk*	*Iron Oxide*
Romanée-Conti	10	30	45	9
Clos de Vougeot	47	36·7	12	3
Nuits St. Georges	63·4	19·2	12·2	4·16
Ermitage	54·6	2	35·7	3·5
Côte-Rôtie (Blonde)	86·5	7·03	3·2	2·1

Dr. Alex Muir found some interesting parallels between a number of English vineyard soils (both existing ones and those in the past) and certain continental areas.[70] The stony Bordeaux *Graves* were similar to certain Hampshire soils around Bournemouth, such as Beaulieu Abbey, and at Castle Hedingham and Raleigh, Essex. Parallels with the brick-earth vineyards of Bordeaux were the Thames valley and Reading gravels, while the chalky downland soils of Dorset, Hampshire, Kent and Sussex were like the Champagne marls.

Vinifera roots grow well in chalky soils, but if grafted vines are being used, a chalk-tolerant stock must be employed.

Manures

It is obvious that plant food taken out of the soil by a crop must be returned to it if production is to continue. In point of fact the actual wine removes very little of these essential plant foods—

nitrogen, phosphate and potash—but the leaves and shoots take a considerable quantity. For an average crop of about 250 gallons per acre Chancrin gives the following figures (converted):

Plant food used per acre			
Food	*Leaves, shoots and pomace* lb.	*Wine* lb.	*Total* lb.
Nitrogen (N)	31·5	less than 3	34·5
Phosphate (P_2O_5)	9	1	10
Potash (K_2O)	31·4	6·6	38

It should be noted that fine wines take considerably more plant food from the soil than do ordinary growths.[20]

Thus, provided the leaves, shoots and press cake are returned to the soil each season very modest demands are being made on it. Less than 800 lb. per acre of compost (instead of a usual dressing of some 8 tons) would supply this; or about 20 lb. per acre of a combined fertilizer, say 15.6.30 (15 N, 6 P_2O_5, 30 K_2O) would give the equivalent of the food taken out by the wine. But in practice one needs to apply rather more food than is taken out by the crop.

The vigneron should compost as much as possible of his residues; burning the prunings is fun but a great waste of plant food. The nitrogen is destroyed and much of the phosphate and potash go off in the smoke to fertilize the neighbours' land. Fire, of course, will destroy a considerable number of the winter stages of the two mildews and *Botrytis*, but a well-made compost will heat sufficiently to do that too. The grower should make compost not so much for the mystique associated with 'organic farming', but because, first, it is so much cheaper than chemical fertilizers and, secondly, the plant can draw on the organic food in the soil, as required over the whole season. The soil structure is also improved. Compost will usually supply most of the trace elements required by the vines, except possibly magnesium in very chalky soils.

Potash is much used by the vine; it occurs as potassium tartrate in wines and a soil analysis on a potential vineyard site is valuable in showing whether this element is present in reasonable or small amounts. If the content is low, a dressing of potash should be given at least every other year. The marc or pomace from the wine press is particularly rich in potash. A vineyard soil low in potash is one containing less than 0·1 per cent of K_2O.

It should be remembered that there is a considerable loss of nitrogen when making compost and that the woody vine prunings resist decomposition for some time. However, as the vine is a long-term crop the addition of a certain amount of undecayed organic matter to the soil does not matter. It will eventually break down and give back its constituent foods to the earth. The shortage of nitrogen must be made up either by applying more compost from some other source, or by using an organic waste or a chemical fertilizer, such as artificial urea, by making a nitrogen-enriched compost in the usual way or using a leguminous cover crop between the rows.

The amount of manure a vineyard needs naturally depends on the soil, site, previous crop and kind of grape being grown. As a general guide a typical three-year programme for a Midi vineyard on good soil,[20] producing about 28 hectolitres per hectare (250 gallons per acre), is given below; these figures are equally appropriate for vineyards elsewhere.

			Per acre. Plant food supplied		
Year	*Manure*	*Quantity per acre*	*Nitrogen* (N)	*Phosphate* (P_2O_5)	*Potash* (K_2O)
			lb.	lb.	lb.
1st	Farmyard manure or compost	8 tons	83	54	93
	Basic slag	270 lb.	—	40	—
	Sulphate of potash	50 lb.	—	—	27
			83	94	120
2nd	Dried blood	90 lb.	9	—	—
	Basic slag	270 lb.	—	40	—
	Sulphate of potash	50 lb.	—	—	27
	Gypsum	270 lb.	—	—	—
3rd	Dried blood	90 lb.	9	—	—
	Hoof and horn	90 lb.	12	—	—
	Basic slag	270 lb.	—	40	—
	Sulphate of potash	50 lb.	—	—	27
	Gypsum	270 lb.	—	—	—
			21	40	27

The organic manures (except dried blood) are applied in winter or early spring and the others in spring and summer. After the third year the programme starts again. On poor soils the quantities would be slightly increased and on chalky soils the gypsum (calcium sulphate) would be omitted.

It is now known that, to grow satisfactorily, most plants need very small quantities of a considerable number of elements—the trace elements—such as boron, zinc, copper, nickel, manganese, iron, magnesium, etc. The majority of soils have these elements in sufficient quantities, or they are supplied by the applications of farmyard manure or compost, but a magnesium deficiency may occur in chalky soils: the calcium somehow prevents the uptake of the element even though it is present in the soil. The leaves are pale and chlorotic with occasional burnt edges. A few pounds of Epsom salts (magnesium sulphate) applied per acre usually remedies this matter.

Vines can be fertilized by means of 'foliar feeds'; while the practice is quite satisfactory, it is usually a troublesome and expensive way of applying nitrogen, though it does make the foliage look nice—'cosmetic fertilizing' in fact.

The manuring programme can have a considerable effect on the control of diseases and this is discussed below (see page 113).

5
Pests

It is unfortunately true that wherever a plant is growing in abundance there also will be found all those forms of life (including man) which can use that plant as food, housing or any other purpose useful to it. When these become too numerous we call them pests. From the vine's point of view man is one of the worst pests it has. Instead of allowing the seeds to be eaten by birds, and scattered far and wide in the droppings, thus extending the plant's habitat, man collects the fruit for himself, subjects it to terrible indignities and finally pushes the seeds, deprived of all their attractive flavours and sugars, on to a compost heap. He may even send them to a factory which will mortify them yet further by extracting the oil. What a life!

However, we must harden our hearts and not start a Society for the Prevention of Cruelty to Vines, for the vine has made itself so attractive to man that the relationship is one of mutual benefit—symbiosis. The world's area of cultivated vines is some 9 million hectares; a greater area is now occupied by the vine than if it had been left alone and allowed to spread naturally.

Vine pests throughout the world embrace viruses and nearly all the forms of life—bacteria, fungi, eelworms, mites, insects, birds and mammals. Britain is fortunate in that several severe pests in continental Europe are not known here, for instance the phylloxera and one of the three grape caterpillars—the 'Pyrale', *Cochylis* and *Eudemis*. Only the last has not been found in Britain, but the other two have not yet been recorded on vines. It is probable that there is a vine-adapted strain in continental Europe which it is most desirable to exclude from Britain. When returning from a visit abroad, especially from a vineyard, never, never bring back a plant cunningly concealed in your spongebag, for you could easily bring in the phylloxera or one or all of the caterpillars, their pupae or eggs, and do much harm to the new industry. Our freedom from these pests is a considerable economic advantage.

Major insect pests (or races of pests) of continental Europe not found in Britain

(i) Phylloxera

The correct scientific name of this insect is *Phylloxera vitifolii* Fitch, though it is usually called *Phylloxera vastatrix* Planchon. It is an aphid, the same family of insects as the greenfly of roses. As well as living on leaves, flowers and stems, aphids can also live on roots. Many gardeners have lost lettuce crops to root aphids and household primulas in pots often die for the same reason. The phylloxera aphid can live on all parts of the plant and it has a life history of incredible complexity. One could easily believe it had been drafted by an income tax legislator. So incredible was it that about a century ago, when the cycle was worked out by a French entomologist (Lichtenstein), he was called the '*romancier du phylloxera*', although he was quite correct in his account of its life. I do not give the life-history here, but refer anyone who wants to know it to another book.[80]

The phylloxera is an American insect which lived on the many species of American vines on both the aerial parts and the roots. The struggle for survival over the millennia had reached a kind of concordat situation, such as might be reached between pope and monarch or cops and robbers. Although some American *Vitaceae* (such as *labrusca*) could be killed by the root form, others were entirely resistant to it (*rotundifolia* and *berlandieri* for example) and none of them was extinguished by the creature.

On the European vine (*vinifera*) the case was very different. The insect did not much like the leaves and lived on the roots. There it injected a poison which slowly and inevitably killed the plant. The insects left the dying plant, sought another one and repeated the process. The pest can be controlled by chemical or biological means, the latter being far more satisfactory. In the first system fumigants are injected into the soil, killing the root forms. Treatment is expensive and must be done annually. The biological method of control is to graft the desirable *viniferas* on to stocks of American species, the stocks not being affected by the root forms of the aphid.

Strangely enough the first record of the insect's presence in Europe was at Hammersmith, London, in 1863, but the pest did not spread from there. Soon an 'unknown disease' was devastating France. It was the phylloxera, of course. Two-thirds of that country's vineyards were destroyed and all of them* were replanted with *viniferas* grafted on to American phylloxera-resistant rootstocks—about 12,000 million plants, an enormous task.

The question of whether the British vigneron should use grafted or own-roots plants has already been discussed (see page 73). Outbreaks of phylloxera may occur from time to time, in fact they have occurred, but vineyards in our country are so distant one from the other that the insect is unlikely to become established permanently. The following outbreaks in the United Kingdom and Eire have been recorded by the British Ministry of Agriculture:

Year	*Place*
1863	Hammersmith, London
1867	Cheshire
	Wicklow, Eire
1868	Elvaston, Scotland
1876	Drumlanrig, Scotland
	Gunnersbury, London
1878	Monmouthshire
	Slough, Berks.
1884	Dorking, Surrey
1890	Fort William, Scotland
1904	Sussex
1907	Kent
1908	Hertfordshire
1911	Berkshire
1912	Gloucestershire
1934	Berkshire
1935	Sunningdale, Berks.
1944	Wentworth, Yorks.
1956	Evesham, Worcs.
	Hampshire

Twenty outbreaks in 110 years is not a very high figure and in all cases the infestation was eliminated, mostly by destroying the vines in question and any of their neighbours'.

In addition to Britain there are a number of other countries

* Except a very few in sandy areas, such as Aigres-Mortes. The insect cannot live in sandy soils.

which still grow own-roots vines. They are Cyprus, Chile, Peru and parts of Portugal and Russia. There are also a few patches of ungrafted and thriving vines in France and Germany. The most remarkable such plot is a hectare of Bollinger vines in the Champagne. Why it remains free from phylloxera is a mystery. Presumably the soil contains enough natural or synthetic insecticide (possibly DDT) to prevent the establishment of the pest. The case of Russia is also interesting. Their *vinifera* rooted vines are attacked by phylloxera and chemical control is practised, mostly the injection of hexachlorobutadiene into the soil every year. Why resistant rootstocks are not more used is difficult to say. Perhaps American roots are ideologically unsound and one can only hope that the present *détente* will lead to this more satisfactory method of control being adopted. On the other hand, perhaps the insecticide treatment is a temporary measure while the grafted replacements are being prepared. A third possibility is that the authorities wish to see the pesticide trade maintained even though it is a more expensive way of pest control than are biological methods.

The vigneron in Britain, particularly the one with own-roots *vinifera* vines, should keep a sharp look-out for phylloxera. The symptoms are usually first noted in the summer. An infected plant will show the normally dark green leaves taking on a reddish colour. By the end of August the leaves will have fallen. The fruit will not develop and will dry up. If dug up at this stage the roots will have swellings and a number of yellow aphids on them. In the winter the shoots will be found to be dry and brittle. That year or the following year the plant will die, having been killed by the aphids feeding on the roots. But remember that a dead plant on being dug up will have no aphids on it because as the plant fails they leave to seek living roots and food, carrying yet more destruction with them. Rotting of the roots by fungi and bacteria follows.

Phylloxera is spread by its winged stages and by the planting of vines with infected roots. All imports from abroad should be carefully inspected to see if there are any swellings on the roots, in which case they are very suspect. The pest can also be found on vine leaves, where it forms small galls with a slit-like opening on the upper side. The insect does not much like *vinifera* leaves but enjoys most of the American species' leaves and those of some of the hybrids. If you have any American *Vitis* species in your garden or vineyard, keep a watch for swellings on the leaves, they could be housing the phylloxera, brought in on the wind. But bear in mind that these swellings can also occur on *vinifera* leaves, though they are not common. On the other hand the galls might have been caused by the vine leaf midge (*Janetiella oenophila*). A close examination, particularly if a hand lens is available, shows the difference between these two kinds of gall. The phylloxera gall is on the underside of the leaf with coarse hairs around it and with an entrance to it from the surface. It contains one or more aphids, a characteristic feature being that they have legs. The midge gall is hard and shiny on the top of the leaf and extends also to the underside, where it is hairy. It contains some tiny legless grubs.

The phylloxera is a notifiable pest; that is, its presence, or suspected presence, must be reported to the police or the Ministry of Agriculture. Remember too that the phylloxera can live on some American roots without doing them any harm and that such roots can be a focus of infection for *vinifera* roots.

If phylloxera is confirmed, the action the Ministry would take has a considerable bearing on the question of planting own-roots or grafted vines. The Ministry might decree that the whole vineyard should be rooted up and burnt or that insecticidal treatment be given until the infected area was healthy again: perhaps the latter course of action would be more likely. If destruction became necessary, there are no provisions for compensating the owner for his loss, as is the case with Colorado beetle and potatoes.

(*ii*) *Caterpillars*

The three serious continental caterpillar pests not found in Britain as pests are:

a. The pyralid moth, *Sparganothis pilleriana* Schiff., has been found in England but not on vines. It measures about an inch across the wings. The forewings are roughly rectangular, pale yellow in colour with wavy red bands, more marked in the male than the female. Although the insect can feed on a large number of plants (such as *Artemisia*, elm, *Galium*, blackberry, ash and strawberry), on the continent it prefers the vine and is only found on the other growths in the neighbourhood of vineyards.

The greenish-yellow eggs are laid in summer, usually during twilight, in roundish or rectangular patches (fifty or sixty eggs) on the upper sides of the leaves. They hatch after about ten days and the young caterpillars seek hiding places in which to pass the winter, such as beneath loose bark and in cracks. The next spring (mid April) the caterpillars come out of their refuges and start eating the young shoots, laying much silk around them. As they grow their appetites increase and they spin more and more leaves together. They first attack the tops of shoots, but having consumed all these they then turn to the flowers and fruit. They can do enormous damage. When fully grown the caterpillar is just over an inch long. It then pupates, and emerges as a moth a fortnight later.

The usual insecticidal treatments are effective against this pest, but another early treatment of a mechanical nature is of interest. Ninety years ago in France the pest was causing great losses. The life-history was worked out by the great entomologist Audouin and it was realized that the winter stage was the vulnerable one. Small portable boilers were wheeled or carried down the vine rows, and the hiding places of the caterpillars scrubbed out with hot water and a wire brush—the *échaudage*. A sort of coffee pot was used to pour on the water! It was very effective.

b. *Eudemis* (*Lobesia botrana* Schiff.) and *Cochylis* (*Eupoecilia ambiguella* Hb.). The caterpillars of these two small *Tortrix* moths can cause immense damage by attacking the young fruit bunches. The fully grown *Eudemis* caterpillars are about ⅓-inch long and green in colour with a shiny black head and thoracic plate. The *Cochylis* caterpillars are a little bigger; they are greenish grey or reddish in colour with lilac reflections from the body.

They are both typical *Tortrix* caterpillars. If touched they wriggle violently (hence the name *Tortrix*) and drop, spinning out a thread as they fall. They ruin the bunches by eating, excreting and spinning silk. These caterpillars can likewise live on other plants—the *Eudemis*, on *Daphne nidium*, hawthorn, black bryony, *Clematis flammula*, ivy, privet, arbutus and sea scilla. The plants on which the *Cochylis* can also live include laurel, hawthorn, dogberry and the spindle-tree.

If small caterpillars are found attacking vine flowers or the young bunches of grapes the matter should immediately be reported to the Ministry of Agriculture. Taken in time, an incipient invasion can be wiped out and much future trouble prevented.

Minor pests of continental Europe, not found as pests in Britain at present, but which could be imported and become established

One of the difficulties of applied entomology is that a more or less harmless insect in one area may become a devastating pest when introduced to another. The phylloxera introduced to Europe and the gipsy moth to North America are examples. The following insects are not found as pests in Britain and are minor pests in continental Europe:

(*i*) *The vine flea-beetle* (Altica lythri *Aubé*)

A typical flea-beetle of a metallic blue-green colour, eating on both sides of the leaves.

(*ii*) *The cigar-maker* (Byctiscus betulae *L.*)

A weevil which rolls a leaf or leaves together something like a tiny cigar.

(*iii*) *The scribbler or writer beetle* (Adoxus obscurus)

This Chrysomalid beetle is so called because it eats off the cuticle of leaves leaving scribble-like marks on them. It is also known as 'the little devil'. It is not at all common in Europe now, but was once a serious pest. Accidentally introduced into California it caused a lot of damage there. The larval stage is spent in the soil beneath the vines and it is thought that it does not like the American roots. The scribbler could thus become a pest in Britain in own-roots vineyards.

Insects reported on vines by Kirby and Spence (1818)[55]

None of the following insects is a serious pest at present. The Reverends William Kirby and William Spence, 'the fathers of entomology' in England in the early and mid nineteenth century, recorded a number of insects on the vine—mostly from abroad—which should just be mentioned for there is always the threat that they might become serious at some future time.

The two scientific clergymen start this account with a reference to the Bible. The Children of Israel were always being threatened with the direst consequences for disobedience. For instance: 'Thou shalt plant vineyards and dress them, but shalt neither drink

of the wine, nor gather the grapes; for the worms shall eat them.' Deuteronomy, xxviii, 39.
The above sounds very like the *Cochylis* and *Eudemis*, which our authors did not find in England, we being more blessed than Israel.

In Hungary the beetle *Lethrus cephalotes* (now known as *L. apterus*) did great damage by cutting off the ends of shoots, and five other beetles, including the vine flea-beetle already mentioned, were noted as doing damage in France, one of them, *Otiorhynchus sulcatus*, the vine weevil, being known in Britain.

The three caterpillar pests were also noted as occurring in France, Germany and Italy. The authors also recorded scale insects, species of *Coccus* as they put it.

Insects and other arthropods now found on vines in Britain

SERIOUS PESTS

There is only one serious insect pest of vines in Britain at the moment—wasps. They do much good in the early part of the year by preying on insects (most of them damaging or a nuisance) such as flies, but later in the season they attack fruit. They can do enormous damage to early-cropping vines. Of the seven species of social wasps in Britain two are the main culprits—the common wasp (*Vespa vulgaris* L.) and the tree wasp (*V. sylvestris* Scop.). The former nests in holes in the ground and the latter in trees and old buildings.

In the late summer and early autumn virgin females (queens) and numerous males are produced. After mating, the queens seek hibernation quarters (frequently in houses) and the workers and males, their tasks completed, indulge in an orgy of feeding for themselves; there is no brood left for them to care for. They are attracted to the autumnal sugars in plants. Soon the cold kills them all. The males have no sting and often accumulate on the late ivy clusters of fruit. A bold man may impress his companions by running his fingers among them and saying 'Wasps don't sting me!', stoically biting his lip if a stray worker happens to be there. Males can be distinguished from workers by the fact that they (the males) are bigger and have longer antennae, thirteen segments instead of twelve, not, I must admit, an easy observation to make *in vivo*.

CONTROL OF WASPS

The best method of wasp control for the vigneron is to grow the late-ripening varieties. In such cases it will be early October before the grapes are attractive to wasps and the vast majority of the insects will have died and the queens have hibernated.

Other methods of control are putting plastic bags around the fruit and destroying nests. A Cambridge vigneron found that applying bags is not as laborious as might be expected and also protected the grapes from their worst pest—birds.

To destroy nests they first have to be located. An idea of where they are can often be obtained by carefully watching the arrival and departure of workers. Another method advocated is to set out a plate of jam to attract the creatures and to tie a piece of white cotton, say nine inches long, to a wasp. This may sound something like the mouse's solution to the problem—putting a bell on the cat! But if a cotton is attached to a wasp, after feeding it flies off home and the thread slows it down so much that one can follow it and get an idea of the direction of the nest, if not find the actual nest itself. A loop or noose on the cotton sometimes allows one to lasso a wasp!

Modern insecticide powders put into a nest mouth, preferably in the evening, soon destroy it. The fluttering of the wings disturbs the powder and it is carried inside.

Wasp candy was a brilliant idea for wasp control that did not work out in practice. Ordinary wasp baits, such as jam containing an insecticide, attract and kill both bees and wasps, but it is most undesirable to kill the former insects. Bees have sucking mouths and wasps biting ones, hence the idea of wasp candy. A hard, poisoned candy could be taken by wasps cutting off bits with their strong jaws, a thing bees could not do; thus the wasps were killed and not the bees. However, rain and dew softened the candy, enabling bees to suck it up as well. Children, frequently a severe pest of ripening grapes, might also take the insecticidal candy, which, though it would be unlikely to harm them, could give rise to considerable consternation and is probably illegal too.

MINOR PESTS

(*i*) *The vine weevil* (Otiorhynchus sulcatus *F.*)

This weevil is occasionally found, particularly on wall vines and in greenhouses. The larvae feed on the crowns of many plants, causing them to wilt or even die. The adult, a largish black weevil, eats notches out of the margins of leaves, not doing much damage.

(ii) Scale insects

Species of *Parthenolecanium* Bouché—soft scales—are sometimes found on vines, again mostly those growing on walls or in greenhouses. Their presence is usually indicated by a procession of ants going to and leaving the vines. Those attendants are attracted by the sugary secretion given off by the insects.

The cottony scale (*Pulvinaria vitis* L.) is usually only found in glasshouses.

(iii) Red spider mites (Panonychus ulmi)

If these mites become very numerous they cause yellowing and hardening of the leaves and do much damage. They can be controlled by the usual insecticidal sprays, but such sprays should not be given unless there is a real threat of damage. If spraying with an insecticide is started it will have to be continued, as the predators living on the mites and keeping down their numbers will also be killed.

(iv) Vine erinosis or felting

This is a condition of the leaf which at first sight may appear to be very alarming but in reality is not. A mite—*Eriophyes vitis* Pgst.—feeding on the undersides of leaves, causes a white felt-like growth to appear, starting in the axils of the veins. The colour gradually changes to reddish as the season advances. It has never been recorded as doing any significant damage.

(v) Sphinx moth caterpillars and others

O. S. Wilson records a number of caterpillars as feeding on vine leaves, but they never seem to have done any damage and mostly have been regarded as entomological rarities and a feather in the cap of any discoverer of one of them.[120]

The names given below complete the list of vine-feeding Lepidoptera:

Striped hawk moth	*Deilephilia lineata* F.
Silver-striped hawk moth	*Choerocampa celerio* L.

Elephant hawk moth	*C. elpenor* L.
Garden dark moth	*Agrotis nigricans* L.
White line dart moth	*A. tritici* L.
Streaked dart moth	*A. aquilina* W.V.
Square spot dart moth	*A. obelisca* W.V.
Garden tiger moth	*Arctia caja* L.
Cream spot tiger moth	*A. villica* L.

(The scientific names are those given by Wilson.[120])

I once found one small caterpillar of the apple-leaf *Tortrix* (*Ditula angustiorana* Haw.) on a vine leaf in Kent.

(*vi*) *Greenfly*

Apart from the phylloxera, a green aphid (*Aphis vitis*) may attack the vine. The infestation is seldom serious.

An anomaly

A recent report from Hungary says that the cabbage moth (*Mamestra brassicae* L.) has been found feeding on vines. In Britain this insect does much damage to Brassica crops but has not been recorded on vines.[47]

Molluscs

Slugs and snails can damage young growths and the latter may attack the ripening fruit. They may be controlled with metaldehyde baits or by hand picking. Vine-fed snails are considered a table luxury in some parts, so perhaps there should be a market for them, though to me on the table snails only taste like bits of garlic-flavoured indiarubber.

Warm-blooded animals

(*i*) *Rabbits and hares*

Rabbits in Britain have returned in sufficient numbers to be a considerable threat to vineyards. Two kinds of damage can be done. The more serious form is the barking of young or established plants and the other is the eating of foliage and grapes during the growing season.

There is only one solution—protective wiring—and the usual method is to wire round the outside, including the headland, of the whole vineyard. The alternative is to put a protective collar of wire around the stem of each vine but the latter method is very laborious and actually uses more wire. An acre is a square of about 70 × 70 yards. A 4-yard headland on two opposite sides of the square gives us a length of wire (2 × 70) + (2 × 78) = 296 yards. The number of collars needed varies, of course, with the planting distance. At 5 × 5 ft 1,742 plants per acre are required. At 15 inches per collar this gives us

$$\frac{1742 \times 15}{36} = 726 \text{ yards of wire.}$$

Under a Lenz Moser system 770 collars per acre would be needed:

$$\frac{770 \times 15}{36} = 321 \text{ yards,}$$

slightly more than wiring round the whole acre, and the collars give very little protection to the foliage. Another point is that the larger the block wired, provided the block is a square or nearly so, the less wire per acre will be used. Thus a 4-acre square block (19,360 sq. yds) would have a side of 139 yards. Allowing for the two headlands the run of wire needed would be (2 × 139) + (2 × 147) = 572 yards, or 143 yards per acre. The nearer the plan of a vineyard approaches a circle the less will be the perimeter wiring for any given area enclosed.

Care must be taken that rabbits do not get into the vineyard by burrowing under the wire, jumping over it or running through a gate left open. The reproductive power of rabbits is notorious. A female will have from five to eight litters a year, each having about six young, and the females start to breed at six months old. They would be only too delighted to be inside a vineyard with a nice fence protecting them from foxes, dogs and stoats! To prevent rabbits burrowing under the wire a better way than sinking the wire into the soil is to curve 9 inches or 1 foot back outwards on the surface of the soil, weighting it down with a turf or clod of earth at intervals. Weed and grass growth will soon fix it. In the soil the wire quickly rots. On the surface it lasts longer and defeats the rabbits. They repeatedly come up against the wire, but when they try to burrow underneath it they are defeated by the mesh every time. They are not very intelligent; they never seem to think of starting their burrow farther back, where, of course, they would be successful. How clever we humans are! As the poet said—'What a piece of work is man! How noble in reason! How infinite in faculty . . .'[98] But, alas, not in everything.

Hares cause nothing like as much damage as rabbits, nor do they breed so rapidly—four litters a year with two to four leverets per litter. Remember that, in addition to your rabbit-proof fence, you must also have rabbit-proof gates and keep them closed even if you are working in the vineyard. A pair can easily slip in when your back is turned.

(ii) Birds

Birds are the worst pest the English vineyards have. The two most damaging genera are blackbirds and thrushes (Turdidae) but many other birds can cause much damage, among them sparrows (*Passer* spp.) and starlings (Sturnidae). Pheasants (*Phasianus*) can also take much fruit from mid August onwards.

Bird control is beset with difficulties, some of which were recently set out in a paper given by Mr. R. J. P. Thearle, who pointed out that the problems could have social, legal and practical aspects.[107] The social consideration is that the public reacts very strongly against any measures resulting in the killing of birds, even when they are severe pests. The legal side is that under the Protection of Birds Act of 1954 all birds are protected, except those listed in a schedule (which includes most of the vineyard pest birds) to the act, which may be killed, but only by an 'authorized person'. To achieve this distinction one has to prove to the satisfaction of the authorities that the scheduled birds are causing losses, and it is not much use becoming an authorized person *after* the loss has occurred.

The practical problem is that birds move long distances and breed rapidly so that action taken by a vigneron over a comparatively small area will make no difference to the occupying population. If, for instance, all the blackbirds in the area of a vineyard were killed, surplus birds from the neighbouring districts would merely move in and repopulate the original district. A bird pair makes a territory and holds it against all invaders seeking to take it over. If the pair dies or is killed, a waiting male from outside immediately adopts the territory, and then attracts a mate, leaving the vineyard as much attacked as ever.

Consequently, bird control in vineyards has to be preventive rather than destructive, and the most satisfactory system is to cover the whole vineyard with netting from mid August until after the harvest. It is expensive—some £300 per acre—but gives the vigneron much peace of mind. Modern plastic netting is long-lasting.

Scaring devices can be used, but are only effective for short periods. Birds soon get used to them and if that system is used the type of scarer must continually be changed. Scarers may be bangers, plastic or aluminium strips waving in the breeze, artificial hawks flown from a small gas balloon, the broadcasting of alarm or distress calls and the all-day patrol of a man armed with a gun, particularly at dawn and evening twilight. The latter, of course, is a seven-day-a-week job. In mid August the individual must be there from at least an hour before sunrise to an hour after sunset, say 3:45 a.m. to 8:30 p.m. Bangers usually work on bottled gas. One particularly effective model shoots a cannister up a mast at the top of which, say 20 feet up, the explosion takes place, thus being heard over a much greater area than if it occurred at ground level, where the sound tends to be masked by the foliage. Bangers, of course, and the broadcasting of alarm calls have to start at dawn and thus give rise to considerable protests from local residents, let alone the vigneron's family.

Gas balloons have to be topped up with hydrogen from time to time, coal gas no longer being available in most areas, and they are liable to carry away in gusty weather. At first the birds are frightened by the menacing hawks flying over the vineyard, but soon get used to them. Perhaps they realize that their dreaded enemies never seem to stoop—they are just paper tigers.

Pheasants (*Phasianus colchicus*) are not protected by the Protection of Birds Act, 1954, and may be shot outside the close season, 2nd February to 30th September, but they can do considerable damage from mid August onwards, when it is an offence to shoot them, that is before the end of September. However, the Ministry of Agriculture is entitled to make an Order, under Section 98 of the Agriculture Act, 1947, allowing pheasants to be shot during the close season if it can be shown that the birds are damaging crops. In practice, of course, by the time such an order had been obtained the damage would already have been done. The vigneron, in a well-stocked game area, might be wise to ask in advance for such an order.

Another aspect of pheasant damage is that the birds are game and belong to the landlord. If you own the land, there is no difficulty about shooting them (in the open season) provided you have the appropriate gun and game licences. It makes no difference that the game flies over from someone else's land. As long ago as the *Bolston* case in 1597 (which was concerned with rabbits, but the same applies to birds), it was held, in the picturesque language of that age, that 'so soon as the conies come on his neighbour's land, he may kill them, for they are *ferae naturae*'. It was held, in the case of *Farrer* v. *Nelson* in 1885, that the owner of shooting rights who kept an unnatural number of game birds on his land, and increased their numbers artificially, was liable for damage to crops as the tenant was not in a position to prevent the destruction by shooting the birds. But in the later case of *Seligman* v. *Docker* (1949), where the number of birds increased at an alarming rate owing to the excellence of the summer of 1947, but did so naturally, and caused damage to crops, it was held that the landlord was not liable for the damage as the increase had been natural and—although the tenant was not allowed to shoot them—the landlord was not under any obligation to shoot them or to reduce their numbers to normal; though it must be added that, in the last-named case, the landlord clearly went out shooting a great deal, and there was no question of his neglecting his shoot. In the circumstances, if anyone takes a lease of agricultural land with a view to growing grapes, he would be well advised specifically to cover the question of shooting game birds that are found attacking the grapes. If he does not, he is unlikely to have any remedy other than the expensive measures of protection that have already been referred to.

Diseases

Next to birds the two mildews are the worst pests of English vineyards. Both of them came from America and, like the phylloxera (see page 88), one of them—the powdery mildew—was first found in Britain.

(*i*) *Powdery mildew or Oidium* (Uncinula necator *Schuw. Burr.*)

At Margate in 1845 Mr. John Tucker, gardener to Sir John Slater, sent some diseased vine leaves to the Reverend M. J. Berkeley, the famous 'fungiferous curate' at King's Cliffe, Northants., and the inventor of the science of 'vegetable pathology'. Berkeley called it *Oidium Tuckeri* in honour of its discoverer.[10] Berkeley saw only the vegetative stages and so named it wrongly, but the disease is still frequently called the *Oidium* in ordinary conversation. The sexual stages were found in 1892 and the fungus was named *Uncinula necator* Burr.

A white powdery mould appears on leaves, stems and fruit; spreading rapidly, it turns purplish as it ages. The mould grows on the surface, pushing 'roots' (haustoria) into the tissues and robbing them of food. Leaves die and the fruit cracks and fails to develop. The fungus passes the winter in two ways, as persistent mycelium on the shoots and as the perfect (sexual) stage—the perithecia, also on the wood. These small black bodies can sometimes be seen in the autumn. Both forms release infective conidia (summer spores) in the spring.

Soon after the Margate discovery the disease was doing much damage in France. A M. Grison, in charge of the royal grapes at Versailles, was much worried by it and started to look for a remedy. In 1848 the grapes became Republican, but still suffered; in 1851 the disease was controlled, not so much because the grapes had now become Imperial, under Louis Napoleon, but because M. Grison had found that spraying with the diluted yellow liquid obtained by boiling lime and sulphur together protected the crop. He had invented lime-sulphur. It was called '*Eau Grison*'.[80]

While gardens and greenhouses could be sprayed using syringes and 'garden engines' it was quite impractical to do this over vast areas of vineyards. A M. Duchâtre found that dusting with fine sulphur was a cure, and it still is to this day. Sulphur-dusting was the first great success in the history of chemical pest control.

Sulphur dust is blown into the flower bunches in early summer. Lodging there, it helps protect the grapes throughout the season, which is not to say that no more treatments are needed. Fungicides are mainly preventative materials, thus new foliage and fruit must be treated so that spores of the disease falling on them cannot germinate. Three or four more dustings during the season need to be given. Dusting with sulphur also helps in the set of the fruit. Presumably it blows the pollen around more effectively than do natural air movements.

Spraying can also be used to control this disease, in which case wettable sulphurs can be used. Spraying is more expensive and takes more time than does dusting, but the effect is longer lasting in wet weather, provided the spray has had time to dry on the plant. In spraying less sulphur per acre is used. For instance dusting needs about 15–20 lb. per acre and spraying 5 or 6 lb. If wettable sulphur costs more than three times as much as dusting sulphur, it is cheaper to use the latter product, but bear in mind that application costs are less for the dust. Wettable sulphurs are essentially ordinary ground or flowers of sulphur to which a detergent has been added; even soap has been used for the purpose.[64] These wettable sulphurs are proprietary compounds and you can make your own, which will be much cheaper, by adding sulphur to water containing a little household detergent. If this is done, a test must first be made on a few vines to see if the mixture causes damage to the leaves.

Messrs Sandoz's wettable sulphur—Thiovit—is much used in France. In this product the particles are between 1 and 6 microns in diameter and thus the smaller particles are said to be too large to enter leaf stomata and cause damage, making the product very safe to use under hot conditions. Damage from sulphur dusts or sprays is very unlikely in the English climate.

Sulphur dusts are either sublimed sulphur (flowers of sulphur) or ground sulphur, but both these products show a tendency to ball. A much more satisfactory dusting material is obtained if 10 per cent of an inert substance, such as china clay, is added. This is not a process the grower should try doing himself, because of the risk of a cloud of sulphur and air around the mixer being ignited by a spark or cigarette and an explosion resulting.

Sulphur treatments are essentially preventives, though there is some curative action as well, and eliminating curative sprays have been sought and found from the earliest times. One of these was permanganate of potash, about 1 lb. per 100 gallons of water to which 2½ lb. of lime might be added.[20] Modern curative sprays are complex organic fungicides, such as dinocap (capryldinitrophenyl-crotonate, Rohm and Haas Ltd), but they are not much used against powdery mildew, sulphur being so effective and cheap.

A new development in chemical control is the systemic pesticide. This is a chemical which when sprayed on to or watered around a plant is absorbed by it and translocated within the tissues to all parts, killing the pests there (insects or fungi) without harming the plant or its crop. Obviously it makes spraying easier and more effective, and equally obviously it must not leave toxic residues in the crop.

A new systemic fungicide against the vine powdery mildew (and related mildews) was announced in 1975.[34] It is Bupirmate (a pyrimidine fungicide, ICI) and has been given the trade name 'Nimrod'. Nimrod the mighty hunter was Noah's grandson and founded Babylon.[38] Presumably the product hunts the fungi as energetically as did Nimrod the beasts opposing the dominance of the human race. It is also interesting to know that our captains of industry still read the Bible. The new fungicide could well simplify control of powdery mildew.

(*ii*) *Downy mildew* (Plasmopara viticola (*Berk. and Curt.*) *Berl and de Toni*)

This American fungus was originally called *Peronospora viticola* de Bary and was first found in Europe at Contras (Lot-et-Garonne) on an American grape ('Jacquez'), by Planchon in 1878. It is somewhat ironic that Planchon found it, because he had issued repeated warnings on the dangers of importing this disease in the rush to secure wood from America for phylloxera-resistant rootstocks. The disease is still commonly referred to by its original generic name—*Peronospora*—just as the powdery mildew is still called *Oidium*.

The disease spread rapidly over Europe; it was recorded in England in 1894, but not again until the 1920s. In the 1880s in France it started to cause enormous losses and was as much feared as the phylloxera.

The disease first causes pale translucent patches ('oil spots') on the upper sides of the leaves which slowly spread. On the undersides a white down can be seen; the leaves wither and drop, in consequence of which the fruit does not develop. The fruit itself can also be affected. It turns brown and does not ripen.

The white down consists of branched stems (conidiophores) bearing spores (conidia) and they usually emerge from the stomata or breathing pores on the underside of a leaf. Oospores (resting bodies) are formed in the autumn, fall to the ground and germinate by means of a short germ tube in the following spring. This tube forms itself into a conidium in which spores, motile in a drop of water (zoospores) are produced. Rain splashes these spores on to the lower leaves where the first infections take place and from whence they infect the whole vineyard.

The more or less accidental discovery of the remedy—Bordeaux mixture—is well known and will only be briefly recorded here. Many vine growers used to splash the grapes alongside roads with verdigris, to discourage pilfering by passers-by. The verdigris was copper acetate and looked like the pigment Paris Green, a dangerous chemical (acetoarsenite of copper), sometimes used in wallpapers, occasionally leading to the death of children who licked them. Verdigris started to become expensive, because it was used for making Paris Green, the demand for which was increasing in America, where it was used as an insecticide. Also the suggestion was growing that the verdigris-splashed vines were as poisonous as if they had actually been treated with Paris Green. The possibility of 'poisoned wine' always creates alarm. It was then found that a mixture of solutions of copper sulphate and milk of lime gave a precipitate which looked very like verdigris when splashed on the leaves and this brew was used alongside a road by M. David, the manager at Château Beaucaillou, Gironde, in 1882. Professor A. Millardet, walking along this road one day, noted that the splashed vines were free of downy mildew, whereas the rest of the vineyard was attacked. The discovery led to the wholesale use of Bordeaux mixture and the vines were saved once again. It is still a cheap and very effective spray.

The original Millardet mixture was very strong, 8 kg of copper sulphate and 15 kg of quicklime in 100 litres of water. The resulting paste could only be applied by flicking it on to the leaves with a brush made of twigs. Nowadays much weaker mixtures are used and the lime is usually purchased as the hydrate. A common formula is 10 lb. of copper sulphate, in the form of snow crystals, put into 50 gallons of water to which, when dissolved, are added 15 lb. of hydrated lime in 50 gallons, making 100 gallons of spray. The formula is expressed as 10.15.100. Note that 100 gallons of water weigh 1,000 lb. The mixture should be made anew each day as it is more effective when fresh, and the copper sulphate must not be dissolved in a metal tank, but in a wooden or plastic one.

Note also that the spray is preventive. The object of spraying is to cover the foliage with a protective layer of fungicide so that spores alighting on it are killed, consequently repeated sprayings throughout the season must be given to protect the new growth. The first treatment should be given when the flower bunch is showing but with the flowers unopened and in a tight bunch. Another spray is given two weeks later, provided the flowers are still unopened, and spraying is continued at fortnightly intervals throughout the season.

Burgundy mixture is the same as Bordeaux mixture but with the 15 lb. of hydrated lime per 100 gallons replaced by 5 lb. of sodium carbonate (washing soda). It was developed in areas (Burgundy) where lime was short.

The literature on these two fungicides is enormous.

There are many modern organic fungicides which substitute for Bordeaux and Burgundy mixtures, such as Antracol (zinc propylenebisdithiocarbamate, made by Bayer, A.G.), Elvaron (dichlofluanid, Bayer), Maneb (manganese ethylene-1,2-bisdithiocarbamate, Du Pont), Propineb (zinc propylenebisdithiocarbamate, Bayer) and Zineb (zinc dithiocarbamate, Du Pont). There is also Sandoz's blunderbuss mixture 'Tri-Miltox' containing copper sulphate, carbonate and oxychloride, mancozeb (zinc and magnesium carbamates), and a 'biological adjuvant, stimulating the activity of the chlorophyll and giving the leaf a dark green colour', the nature of which is not specified, but sounds as if it might be a foliar feed. The product is much used in France.

The organic fungicides are less damaging to the leaf than Bordeaux and Burgundy mixtures, but whether they are worth the extra cost is a moot point which only a comparison of prices can show. If pound for pound an organic fungicide has three times the efficiency of copper sulphate then it is worth at least three times the price of the simple chemical, or even a little more if it is less damaging to the foliage. But it is doubtful if it is worth say five times as much as the copper salt. Freshly made Bordeaux and Burgundy mixtures are cheap and effective: they may not leave the foliage *looking* so attractive, but is the cosmetic effect of a new fungicide worth paying for? Perhaps it is, if the price is not too high. There is a great satisfaction in seeing a fine foliage. On the other hand I find a copper-splashed vine very satisfying because it is the physical evidence of having defeated millions of fungus spores trying to take the crop.

(*iii*) *The Grey Mould* (Botrytis cinerea *Pers*)

This fungus can also cause considerable damage to vineyards, and its attacks on British outdoor vines seem to be increasing. It is a widespread fungus and can be both saprophytic and parasitic—that is it can live on both dead and living plant tissues.

Botrytis cinerea is said to be the conidial form of *Sclerotinia fuckeliana* de Bary, the vine sclerotinia.

In the saprophytic stage the fungus lives on dead wood, old tendrils, leaves, stems and other dead vegetable tissues. Under humid and warmish conditions the fungus pushes out the little cup-like growths called apothecia, a growth of conidiophores bearing bunches of grey spores. They arise from the sclerotia (a resting stage, being a mass of mycelium). The conidial spores (or conidia) produced can infect weakened living tissues, such as yellowing leaves, or they gain entrance into plants through wounds or pruning scars, where the fungus then takes up a parasitic existence. The greyish mould proliferates over leaves, stems and fruit under humid conditions. In due course the overwintering forms are produced—the sclerotia already mentioned and little black flask-like bodies, called perethecia, usually sunk into the tissues of the plant, such as the ripening wood.

Late attacks of *Botrytis cinerea* on the grapes themselves are welcomed in some areas and are known as the *pourriture noble*. At this stage the fungus extracts water from the grapes, increases the sugar concentration of the must and by other reactions also increases the flavour of the wine. For such wines as high grade Sauternes, Château d'Yquem for instance, and the German Beerenauslese and Trockenbeerenauslese, wines of incomparable flavour, only the mouldy grapes are picked and a vineyard may have to be picked over some eight times before the crop is cleared. No trace of mouldy flavour enters the wine.

Early attacks of *Botrytis* on the grapes can be damaging as they stop the development of the berries. Moreover a green *Penicillium* fungus (*P. crustaceum*) may accompany the grey mould and add to the damage. *Botrytis* on the early grapes also transmits a mouldy taste to the wine. We thus have a perfect Janus of a fungus with both good and bad aspects.

Measures against *Botrytis* may be cultural or chemical. The fungus thrives on humidity, consequently wider spaced and airy sites are less likely to be attacked than close planted and very sheltered vineyards. Good natural drainage is also advantageous. Pruning off all dead wood and snags, such as old tendrils, reduces sources of infection and if *Botrytis* has been bad it is better to burn prunings rather than to compost them. An inspection of the prunings with a hand lens will show if the little black perithecia are buried in the wood or not. Comparatively harmless fungi sometimes make dark spots on the surface of shoots and these should not be mistaken for the *Botrytis* perithecia. A pin or needle may be used to dig around such a spot, if the stain goes deep or a tiny flask-shaped body can be dug out then that is a perithecium of the *Botrytis*.

A range of sprays—mostly organic fungicides—is now available for *Botrytis* control. The treatment starts with a winter spray of sodium arsenite or dinitro-ortho cresol. The first is cheap and effective but also very poisonous and might be difficult to acquire. Brewers and winemakers are also extremely susceptible to rumours of their products being poisoned, so that an arsenical winter spray, which would not in any way affect the wine, might give the user's vintage undesirable local publicity.

DNOC (dinitro-ortho-cresol) is also poisonous but can be obtained under the Part II Schedule of the Agriculture (Poisonous Substances) Act, 1952, which requires protective clothing to be worn when it is used. The early sprays used were mixtures of copper and sulphides and were said to be effective, but today the organic fungicides rule this field. Among them may be mentioned:

Benomyl (a carbamate), Benzimidazole, Captafol (also called Difoltan, a hydrophthalimide), Captan (another hydrophthalimide), Dichlorfluanid (a phenylsulphamide, given the trade names of 'Elvaron' and 'Euparen' by Messrs Bayer, A.G.), Fentin hydroxide (triphenyltin hydroxide, also known as 'Du-Ter'), Folpet (a phtalimide), 'Nospore' (a Swiss proprietary mixture of thiophenate, folpet and copper), Oxythioquinox, and thiophenates.

The question next arises whether these chemicals in any way affect the wine made from grapes sprayed with them. Some recent Swiss work indicated that benomyl and thiophenate residues in must and wine up to the tolerance of 3 ppm did not affect fermentation and could be reduced in the wine by adding 100 g of bentonite per hectolitre (1 lb. per 100 gallons) but this treatment spoiled the quality of the wine.[39] It was thought that spraying should stop after mid August. Similar work in the Argentine vineyards showed that late applications of orgno-tin compounds, dichlorfluanid and oxythioquinox affected the taste of the wine.[91]

It is, of course, the late sprayings that are needed mostly in

Botrytis control and so far no adverse effects on the wine in Britain have been noted.

It is also said (in 1904) that *Botrytis* is less serious in own-roots vines than in grafted vines, and it would be interesting to know if this is still the case today in Britain.

Minor troubles

There are a large number of fungi and viruses which can attack vines. Viala alone lists twenty-five fungi in addition to the two mildews.[112] There is also a bacterial pest. New viruses on vines are constantly being recorded and there is at least one physiological condition—*coulure*—which can cause trouble. Three of these fungi will first be noted:

(*i*) *Dead arm* (Phomopsis viticola)

This disease has become more important over the last five years. The causal fungus passes the winter in little black bodies, called pycnidia, scattered over the surface of the wood. As in the case of *Botrytis*, the vigneron must not confuse a certain amount of surface spotting with *Phomopsis* infections. In the autumn and winter peel back the surface layer of spotted wood to expose the green cambium beneath. If the surface stain goes into the cambium the shoot could be bearing dead-arm pycnidia, but if that layer is unmarked then the surface spotting is of no significance.

In the spring the pycnidia shoot out summer spores which are splashed around by rain and infect the new growth and flower clusters. Oval, brownish, dead spots appear at the base of the shoots, which droop and may then die, hence the name 'dead-arm'. Less attacked shoots may be girdled at the base and become brittle and easily broken off by wind or in working the vineyard.

Once again control measures are in two parts—cultural and chemical. Obviously cuttings, stocks and scions for grafting should only be taken from healthy wood. Reduction of humidity by wide spacing, drainage and reduction of foliage reduces the attack. If an examination shows that the prunings are infected, then it will be better to burn them rather than to make them into compost. In England a successful spraying programme is an early spring treatment with dithane with further treatments during the season. In France mancozeb has been used successfully, and in Germany benomyl.

Some growers also give a winter spray of dinitro-orthocresol, a troublesome chemical to use, as already noted.

(*ii*) *Black rot* (Guignardia bidwelli)

This fungus causes black patches on young leaves and later on the fruit. It overwinters as pycnidia, perithecia and sclerotia on prunings and dead tissues. Pycnidia and perithecia are very similar. They are both little black flask-like bodies and differ only in the kind of spores held within them. Sclerotia are little lumps of mycelium.

The usual treatments against downy mildew also control this disease.

(*iii*) *Anthracnose* (Gleosporium ampelophagum, *Sacc.*)

Symptoms of this disease can occur on the fruit, leaves and young vine stems as ash-grey, irregularly shaped, sunken spots, often with a reddish band round the edge.

Control measures consist of winter sprays to destroy over-wintering mycelium. The older writers recommended winter treatment with 10 per cent sulphuric acid to which sulphate of iron was sometimes added.[20] In those days it was an almost impossible material to handle, as the solution attacked all the metal parts of the spraying machine. Today many knapsack sprayers are of plastic and could be used for this sulphuric acid treatment. Of course such a sprayer must have *no* metal parts (unless of gold, platinum or stainless steel!) in contact with the spray fluid. Anyone using such a spray must wear an eyeshield and protective clothing.

The usual anti-downy mildew sprays also help in anthracnose control.

(*iv*) *Viruses*

A number of viruses have been reported from vines, the most serious one being fan leaf (also called golden flavescence and *court noué*). Other viruses named from vines are: corky bark, leafroll, *legno riccio* (rugose wood), linear pattern vein-clearing, mosaic, stem-pitting and yellow speckle. Obviously one would not propagate from virus-infected plants; such plants should be destroyed by burning. In the ordinary way plants attacked by a virus cannot be cured.

(*v*) *Coulure*

This is the condition in which the tiny grapes do not grow out. It is thought to be due largely to lack of pollination of the flowers, though it may also be associated with virus attack. It can sometimes be overcome by increasing phosphate and potash manuring. Cold, wet weather can accentuate the trouble.

6

Integrated pest control

A full programme of chemical pest control is very expensive and is likely to become more so because the process is not a hundred per cent effective. The survivors of such treatments usually have a built-in (genetic) resistance to the chemicals used and so, generation after generation (and the generations among insects and fungi follow each other very rapidly), more and more resistance to that spray (and to chemicals related to it), is set up until finally the product, wonderful as it was at first, is no longer of much use. Then a new chemical must be found, and it is usually a more expensive one.

The alternative to chemical control is biological control, which is almost costless to the vigneron, and consists of using resistant varieties and certain cultural methods against diseases or parasites and predators against animal pests, such as foxes to control rabbits.

Monsieur F. Chaboussou recently pointed to the distinction between the above two methods of pest control by quoting a remark of P. Grison's: 'The biologist cannot long be satisfied with the solution to a problem which merely attacks the trouble itself and not its cause.'[19, 41] Chemical control attacks the problem directly: biological control seeks to attack the cause.

A combination of the two methods is known as *integrated control*. Because of the high cost of purely chemical control, vignerons in continental Europe are becoming increasingly interested in the combined method and the system might help the British grower too, starting, as he does, without several of the mainland pests.

The wholesale preventive spraying of plants is not necessary, because, as J. Grainger has pointed out, a fungus can only grow in them when conditions are right for it.[40] If the conditions are not right, it is a waste of money to spray. Grainger, in Scotland, thought this could be ascertained by measuring a ratio he called C_p/R_s—the weight of total carbohydrate in a shoot and the residual (carbohydrate-free) dry weight of the shoot. Roughly speaking, fungi will not attack below a C_p/R_s ratio of 0:5 and severe attack will occur at ratios above 1:0. The reason the fungus will not develop within the plant at low ratios is because there is no spare food for it in the plant tissues. These are not figures the average vigneron can estimate for himself (though the estimation is not so very difficult) but they could be obtained by research stations. It must also be noted that Grainger was not dealing with vines, but one would expect the same general principle to hold.

Chaboussou, concerned with vines, noted the same point, quoted Grainger, and pointed out that many things, in fact almost anything you did to a vine, altered its physiology and thus its propensity to disease. For instance, the variety 'Folle Blanche' put on to American rootstocks became so vigorous and subject to powdery mildew that it was almost impossible to grow in 1904.[19]

Zineb- and maneb-treated vines gave better crops than the copper-treated ones, but M. Gartel thought the improvement was due to the sprays providing zinc and magnesium. As the sprays made the vine more sensitive to *Botrytis* and powdery mildew it would be better and cheaper to apply the trace elements as small amounts of zinc sulphate and Epsom salts. The organic sprays left the leaf looking very beautiful but were not very successful towards the end of the season when copper often had to be used.

Spraying fertilizers leading to an increase of non-protein nitrogen in the leaves gave rise to more disease and could result in virus multiplication because viruses need simple forms of that element. Consequently vines needed very careful fertilizing and foliar analysis might well become an important element in vine culture and pest control. Analysis of the leaves would show just what the plant needed. Finally Chaboussou quoted an interesting point on the carry-through of organic fungicides to the wine. At a blind tasting a skilled taster put a number of wines into certain groups; it then turned out that the separate groups had each been sprayed in July/August with a different organic fungicide.

The only moral for the British vigneron is not to over-fertilize with nitrogen, to watch for zinc and magnesium deficiency and not use too much organic fungicide, but what the right amounts of both are is difficult to say. The vigneron should also seek more information on this subject, bearing in mind that foliar analysis may be an important tool in the future.

7
Weeds

Weeds have always been a problem for the farmer and vine grower. One of the main features of farm operations—ploughing and harrowing—has been the destruction of weeds. Weeds compete for plant food and water in the soil; strongly growing weeds tend to exude an inhibitory substance depressing the growth of other plants. This seems particularly to be the case with couch grass and the vine; it appears to suffer very much from comparatively small quantities of this weed growing around it. A strong weed growth, of course, cuts off light and air from the vine.

A perfectly weed-free vineyard may delight the vigneron's eye and arouse the admiration of his visitors, but the cost of achieving it may not really be worth while. Provided the main stems (stocks) of the vine are not choked by weed a light growth between the rows can be beneficial. It protects the soil from heavy rain, prevents erosion and violent water run-off. In fact it has the effect of a sward (see page 117). Weeds can also take up nitrates in the soil, which would be lost to drainage water in a weed-free clear culture. This nitrogen is returned to the land as the weed foliage and roots eventually decay. We need to avoid being made over-anxious by the presence of weeds though it is hardly necessary to say that some measure of weed control must be undertaken.

The problem is in two parts: weed control in the row itself and that in the main area between the rows where a sward or a bare fallow may be maintained, or a light growth of weed allowed.

A clover crop between the rows, or a grass sward with a good proportion of clover in it, will supply nitrogen. As is well-known, most leguminous crops carry bacteria which fix the nitrogen of the air into plant food forms and thus save the purchase of nitrogenous manures. There consequently is much to be said for a clover sward between the rows as a means of weed control: it should be mown regularly and the mowings left *in situ*. It is much cheaper than any other management process. Its disadvantages are that it will slightly delay soil heating in spring (see page 67) and that it may increase humidity around the aerial parts of the vines as the sward will be extracting considerable amounts of water from the soil, exhaling them into the atmosphere. However, this removal of soil water can be an advantage, as it forces the vine roots to go deep in search of water, where they usually find potash, much needed by *Vitis* spp. In wet seasons the sward helps remove surplus water. The fruit is usually cleaner in sward vineyards, as less mud is splashed about by heavy rains. The increase of humidity around the vines can encourage disease, but not to any great extent unless the vines are very close.

The alternative to a sward is a bare fallow obtained by (i) cultivation, (ii) use of weedkiller, or (iii) covering the soil with black plastic.

(i) Cultivation is expensive if special machinery has to be bought for any area of less than about 10 acres. On a farm, where cultivators and tractors are available and the rows are wide enough apart to allow of machine passage, the cost is reasonable if it is regarded as marginal to their main use on the farm. Cultivating destroys the surface-feeding roots which, though they supply a lot of plant food, may not much matter, as the vine is deep-rooting. The bare fallow soil surface, especially if stony, reflects a certain amount of light and heat to low-hanging grapes. It also allows them to become splashed with mud in heavy rains. The advantages and disadvantages of any method are an interlocking puzzle all the time. Cultivating in the row can be done with a hand hoe, which is laborious, but effective. There is also an ingenious German machine which cuts the weeds in the row. It is attached to a rotavator, and its spring-loaded blades fall back when they hit a post or vinestock. Obviously it can only be used in a well-established vineyard with stout vinestocks.

(ii) Weedkillers are of two kinds—contact and selective. Contact weedkillers, such as sodium chlorate, kill all plant growth (leaves, stems and roots) that they touch, but some contact weedkillers, such as paraquat and iron sulphate, only kill the green parts of plants and do not affect browned stems and bark. Paraquat, once it reaches the soil, has the great advantage of decomposing into substances harmless to roots. Deep-rooted weeds, such as couch grass, docks and bindweed, may sprout again after paraquat treatment and if they become troublesome the growth must be sprayed once more. Bindweed is particularly persistent in this respect; a slight secondary or tertiary growth can make a vineyard look untidy but in all probability is not doing much harm. Other contact weedkillers are DNOC (see page 108) and iron sulphate, the former being very poisonous and the latter, used in a saturated solution (15%), being cheap and effective, particularly against chickweed (*Stellaria media*). All-plastic machines should be used to spray it because the salt will corrode metal.[20]

The selective herbicides are more active against weeds than crop and there is a very large number of these chemicals. The vine is very sensitive to some selective weedkillers, such as 2,4-D, much used in cereals, and considerable damage can be done to vineyards by the drift of this type of weedkiller from nearby wheat fields. What may appear to be alarming symptoms of fan leaf virus could just be the effect of weedkiller drift from a neighbour's or your own herbicide spraying. However, 2,4-D can be used on vines late in the season if suitable precautions are taken.

A German (Bayer) weedkiller programme for vines is:

February-April: Before bud movement. Casaron G granules (diclobenil), 80–100 lb. per acre.
Do not use on vines of less than three years old and do not use for more than two consecutive seasons.

May: Ustine Special, 9 lb. per acre.
Do not use during blossoming.

July-August: Ustinex KR, 9 lb. per acre.

August-September: Hedonal M, 1·5–3 pints per acre.

September: Hedonal MCPA, 3 pints per acre if broad-leaved weeds are troublesome.

An English programme is:

Late February: Casaron G granules, 80–100 lb. per acre.

Mid March: Gramoxone, 3 pints per sprayed acre. Simazine (triazine), 2–2·5 lb. per sprayed acre.

June/July: Gramoxone, 3 pints per sprayed acre.
Use a shield to protect vine foliage and bunches.

Mid August: MCPA (methylphenoxyacetic acid), for any regrowth of broad-leaved weeds. 3 pints per sprayed acre.

The above programme would cost about £15 per sprayed acre

for chemicals alone. MCPA is similar to 2,4-D and one needs to be very careful when using it on vines.

In Austria Dr. Lenz Moser was against the use of weedkillers in vineyards because he thought that after continual employment they affected the vines. By contrast, in France, M. Juillard concluded that extensive testing had shown weedkillers to be safe if used sensibly.[53] Weedkilling should only be done if the weeds had, or were likely to have, enough growth really to do damage: in short, he was against 'cosmetic' weedkilling. M. Juillard thought it desirable to preserve the indigenous weed flora growing in winter and early spring. It helped maintain soil structure and the local fauna, whose interests should not be neglected (presumably because it could help in pest control). Though swards, whether of weeds or other herbage, were still in the experimental stage, they much reduced costs and soil erosion. Bindweed and couch grass should be tackled in their early stages, for if they became established they were expensive to eliminate. M. Juillard says: 'Thought-out and moderate use of herbicides allows one to maintain soil fertility and to reduce soil erosion. In fact, chemical weedkilling, far from sterilizing the soil, has permitted the development of a less brutal technique than that of mechanical soil working.' Note that the weedkilling must be 'thought-out and moderate'.

Another point to be noted is that as long ago as 1957, FAO (the Food and Agriculture Organization of the United Nations) was advocating chemical weed control for vineyards—dinoseb, pentachlorphenol, monuron (CMU) and chlorpropham.[46] They also stressed the dangerous nature of 2,4-D.

A new weedkiller useful in vineyards for the control of bindweeds (*Convolvulus arvensis* and *C. sepium*) has recently been announced—methazole (Velsicol Corporation). It is active on foliage and lodges only in the surface layer of the soil where it continues to kill seedling weeds for some twelve weeks after application. As most of the vine roots are deep, it thus shows a selective action in vineyards. The rate of application is from 2 to 4 lb. per acre. It is said that after several years' use bindweed is pretty well banished from the vineyard. Killing weeds with sprays is an easy operation and encourages surface rooting of the vine. It is usually cheaper than mechanical cultivation and hand-weeding.

Weedkillers, particularly paraquat, can be used in the row and cultivators employed to keep down weeds in the main inter-row area. This common usage is the reason rates are quoted as 'per sprayed acre'. Using this method, the area occupied by the row must be calculated. For instance, if the sprayed band is 1·5 feet wide then an acre is the number of square feet in an acre (43,560) divided by 1·5 = 29,040 feet = 9,680 yards = 5·5 miles. If the band is 2 feet wide the figure is:

$$\frac{43{,}560}{2 \times 3 \times 1760} = 4{\cdot}125 \text{ miles.}$$

These distances require about 3 pints of a paraquat weedkiller—a very small amount.

Knapsack spraying machines used for weedkillers should never be used for anything else. It is difficult to remove all traces of herbicides from them.

(iii) Covering the whole soil area between the vine rows with black plastic is not a practical proposition, nor is it desirable. It greatly interferes with the physiology of the soil, particularly as regards precipitation. Most of the area gets too little rain, but certain run-off points from the plastic get too much and become water-logged. Perforated plastic cannot be used because weeds merely sprout from the perforations. On the other hand the use of a band of black plastic along the row is a good method of weed control in the row. Naturally it is somewhat difficult to fix in position. T-shaped slits have to be made to accommodate posts and vinestocks, and weeds tend to sprout through these openings. Naturally they will also sprout from any tears or holes in the plastic. With a bit of luck in-the-row plastic will last for two years. It is best used with the sward cover system, because with a bare fallow the sheet tends to get caught in the cultivator tines and be torn away. The plastic has to be weighted down with stones. Clods of earth are not very satisfactory because weeds grow in them.

Whilst writing this book an interesting article by J. F. Roques on weed-killing in French vineyards has appeared in *Outlook on Agriculture*[94A] The author discusses the three main methods of weed control (in fact the ones mentioned above): (a) Traditional cultivations, (b) In-the-row herbicide treatment and cultivations in the inter-row spaces, and (c) Over-all herbicides. M. Roques points out that in (a) the surface roots are destroyed so that the vine does not feed on the surface soil; also that earthing up vines to destroy the in-the-row weeds often damages the plant and the method is labour-intensive and thus expensive. The (b) treatment is cheaper because it eliminates the earthing-up or expensive hand-hoeing in the row, but, since the inter-row spaces are cultivated, it means the vigneron must maintain both spraying and cultivating equipment. The (c) over-all herbicide treatment means that the soil is left undisturbed—an entirely new situation for vineyards. It appears to be advantageous, except in some saline soils (because salts may accumulate), for the roots are able to exploit the rich surface layers of soil. Untilled vineyards also showed greater resistance to spring frosts, and in one trial the no-tillage plots had 56 per cent more grapes at harvest.

Ploughing and cultivating aerate the soil but in doing so kill a lot of earthworms. In no-tillage vineyards there are many more worms present and the passages they make and the wormcasts sent up provide all the aeration needed. Also, roots of weeds killed by the herbicide decay and open up further air passages in the soil. The all-over spraying was a comparatively cheap method of weed control. M. Roques thought that the risk of damaging the vines by accumulation of herbicide in the soil was slight provided the chemicals were used at the recommended rates. In this he differs from Dr. Lenz Moser (see page 117). The grower interested in this controversy should certainly read M. Roques's paper.

8

Wine making

The grapes having been grown, they must be harvested and made into wine.

When one reads the directions for a new card game they sound too impossibly complicated to attempt—but as soon as play starts it all turns out to be fairly straightforward.

So it is with wine making. In this chapter I have gone into the process in some detail, but the less experienced grower might find it less intimidating to begin with the brief summary on page 138.

In England picking the grapes usually takes place in the first fortnight of October. It is a difficult matter to decide on the date because, with good weather, the longer the grapes stay on the vine the more sugar and flavour will they have; but if the weather turns wet the crop can be ruined.

With one eye on his crop and the other (and an ear) on the weather forecasts the grower charts the progress of his grapes towards perfection. His main guide is the sugar content, which can be estimated in three ways: by a refractometer, by calculating the specific gravity, or by chemical analysis. The refractometer bends a ray of polarized light to a greater or lesser extent according to the amount of sugar present. It is a neat but expensive instrument and only tests the juice from a single grape. Obviously many grapes must be taken and the readings pooled. I prefer to get the gravity with a hydrometer, as a much more representative sample can be obtained and it is much cheaper. The specific gravity is the relative weight of a volume of must (juice) compared with the same volume of water at a standard temperature. For instance, if the specific gravity is 1·05 this means that a gallon of must would weigh 10·5 lb. compared with a gallon of water, which weighs 10 lb. From this figure the amount of sugar in the must can be obtained and from that the likely amount of alcohol in the wine. A sample of about ½ lb. of grapes is collected, the juice squeezed out in a small hand press (an aluminium orange squeezer is very good). It is strained and then put into a small glass cylinder and a suitable hydrometer floated in it. The instrument has a graduated stem and a reading is taken at the point where it emerges from the liquid. The higher it floats the more sugar there is in the must.

Hydrometer stems may be marked with a variety of scales, such as specific gravity or Baumé, Twaddle or Oeschler degrees or even in potential alcohol content of the finished wine. Wine making in Britain now closely follows the Germans' practice and they use Oeschler degrees, so we will only use that system. It is directly related to specific gravity. In fact Herr Oeschler's system is the specific gravity to three places of decimals without the decimal point and the figures to the left of the decimal point. 1·05 specific gravity is 50 degrees Oeschler: 1·075 specific gravity is 75 degrees Oeschler. The two figures mean that the musts have respectively 103 and 170 grammes of sugar per litre and a potential alcohol content in the wine of 6 and 10 per cent, or degrees as it is sometimes expressed. This is the French and British practice; the Germans usually express the alcohol content of their wines as grammes (g) per litre. Since alcohol is lighter than water this has the effect of making the German wines appear to be weaker than, in fact, they are. Thus a wine with 11 per cent (degrees) of alcohol by volume has 8·82 per cent of alcohol by weight or 88·2 g of alcohol per litre. The German's 9 per cent wine is the Frenchman's 11 per cent, in general terms.

The advantage of the Oeschler system, to the Germans, is that over the middle range of gravities the degrees Oeschler are the same as grammes of alcohol per litre in the finished wine. Thus if a must has 80° Oe the resulting wine should have 80 g of alcohol per litre.

Chemical analysis to determine sugar is usually only done where very exact figures are required, for instance in estimating the amount of residual sugar in wine. Of course, it can also be used on must. A. Massel's book describes a simplified method of chemical analysis, using Fehling's solution.[66]

Besides sugar the must contains acids, dirt, bits of pulp, skin, pips and so forth, so that in estimating sugar it is usual to subtract one or two degrees from the Oeschler (henceforward referred to as °Oe) figure to allow for them. Reference is then made to a table to see how much sugar there is in the must and what the resulting potential alcohol figure in the wine will be. A table showing this relationship is given in Appendix II.

Enrichment

In northern vineyards the must frequently will not have enough sugar in it, in which case 'enrichment' may take place; that is, sugar is added to the must. In general musts below 70° Oe (9·1 per cent alcohol) will need enrichment. EEC regulations and the new German wine law limit the amount of strengthening that may take place. The regulations are complicated, but roughly speaking, sugar may not add more than 30 g of alcohol per litre (3·8 degrees of alcohol by volume) for white wine and 35 g per litre for reds (equal to adding 4·4 degrees of alcohol by volume). Thus a must having 50° Oe would only produce 6 degrees (6 per cent by volume) in the wine. If it were a white wine it could have sugar added to give it 6 + 3·8 = 9·8 degrees, and if red it could have enough sugar to give it 6 + 4·4 = 10·4 degrees—totals of 9·8 per cent and 10·4 per cent by volume.

The vigneron must also estimate the acidity of the must, which can be done either by tasting it or by chemical analysis. Obviously the latter method is more accurate.

Acidity

Although acidity is a natural and desirable quality in wines it can sometimes be excessive in those from more northern countries. The acids in wine are mostly tartaric, malic and tannic; they are most important in the development of bouquet and quality in the final wines. It is a common experience that as fruit ripens the acidity is reduced, but, relying on taste alone, this reduction may not be as great as it appears to be to the tongue. The human palate regards sweetness and acidity as opposites, when, chemically speaking, they are not. If the rhubarb is too sour we sprinkle more sugar on it and find it delicious, but in fact the acidity has in no way been reduced. A sweet wine may in fact have a higher acidity than a dry wine but appear to the consumer to be less acid. It is the high acidities in the exquisite Trockenbeerenauslese wines that help develop their qualities.

Lack of acidity is seldom a problem in northern vineyards, but in an average or bad season too much acidity in the must can easily occur. Acidity in must or wine is usually expressed as so many grammes per litre, the French tending to calculate the acidity as sulphuric acid (it is not, of course, present as such) and the Germans expressing the results as tartaric acid, which is present in wine. We will use the latter method. The sulphuric acid figure may be converted to the tartaric acid one by multiplying by 1·53 and the reverse process achieved by dividing the tartaric figure by the same factor. Good German musts have from 8 to 12 g per litre and medium ones up to 14 g. Acid is lost in the process of wine making and in general a wine has about three quarters of the acidity of the must from which it was derived.

Acidity is usually estimated by titration with an alkali solution, using litmus paper as the indicator (red = acid; blue = alkaline). Any known strength alkali may be used, such as decinormal solution, but it is convenient to have one giving a direct reading. The strength of the solution should be such that, on a 10-cc (cubic centimetre) sample of wine, the number of cubic centimetres needed just to turn the paper from red to blue will be the number of grammes per litre of tartaric acid in the sample. Thus if the 10 cc of must need 10·4 cc to neutralize it, the sample has 10·4 g of tartaric acid per litre. Such a test solution may be prepared by dissolving 7·47 g of pure potassium hydroxide in a litre of rain or distilled water. The more alcohol plus unfermented sugar a wine has, the higher the acidity should be to make the best of the vintage, and a calculation of the acidity at which to aim can be made on the must. The best acid/sugar ratio is given by Nègre and Françot as being between 1:30 and 1:35, when the acidity is expressed as sulphuric acid.[75] Using tartaric acid the figures become 1:20 to 1:23. Thus a must having 146 g per litre of sugar wants from

$$\frac{146}{20} \text{ to } \frac{146}{23} \text{ g of acidity per litre,}$$

that is from 7·3 to 6·3 g per litre. As may be seen from the table (page 158) the above must would give a low alcohol wine, namely 8·6 per cent. By contrast a rich must, say 220 g per litre of sugar (giving 12·9 per cent alcohol), would want 11 to 9·6 g per litre of acid. It must be remembered, as already mentioned, that the acidity is automatically reduced during the vinification process and during ageing of the wine itself. Musts from northern vineyards are sometimes low in sugar and high in acidity (up to 30 g per litre can be found in bad years). The must can be enriched with sugar or concentrated must (of which more below) and there are various ways of reducing acidity—by adding water and by means of potassium monotartrate or an alkali such as calcium or potassium carbonate.

In my view the addition of water is a most undesirable procedure: it is usually a matter of further diluting an already poor must. The practice is frowned upon by the EEC and it is to be hoped that English vignerons will not use water for this purpose. Where it is done the enriching sugar is added as syrup; it used to be done in Germany but is no longer permitted. No details of this process will be given here; if needed they can be found in the literature.[66]

The commonest method of deacidification is the use of calcium carbonate (chalk), particularly a proprietary product called 'Acidex'. This is fine calcium carbonate mixed with small amounts of the double calcium salt of dextro-tartrate and laevo-malate.

1 g of acidity per litre is removed from 100 litres of must by the addition of 66 g of calcium carbonate.

Avoirdupois equivalents are: 1 g per litre of acidity is removed from 100 gallons of must by the addition of $\frac{2}{3}$ lb. of calcium carbonate. If one wishes to remove, say 5 g of acidity then 5×66 g $= 330$ g of calcium carbonate per 100 litres of wine must be used (equal to $3\frac{1}{3}$ lb. per 100 gallons of wine). Powdered precipitated chalk, B.P. grade, can be purchased from chemists. Tables for the use of 'Acidex' will be found in A. Massel's book.[66] This product is used at about the same rate as calcium carbonate. Treatment with carbonates naturally releases a considerable volume of CO_2 gas. While this makes the must more difficult to handle it does have the advantage of preventing oxidation. Mr. P. A. Hallgarten indicates that deacidification with calcium carbonate leaves a calcium taste in the wine, which is my own experience too, and I think if one is going to reduce acidity with a carbonate it is better to use potassium carbonate.[44, 78] Treatment with calcium carbonate gives a precipitate of calcium tartrate, whereas the use of the alternate potassium salt gives potassium tartrate, a substance quite natural to wine. As the atomic weights of calcium and potassium are very close to each other (40·8 and 39·1) the amounts calculated for calcium carbonate will also serve for potassium carbonate. The only objection to the potash salt is that it is more expensive than the calcium one.

A third method of deacidification is the use of mono-potassium tartrate. It is the one I prefer because—as mentioned above—potassium tartrate is a constituent of natural wine. The mono-tartrate takes up an additional molecule of tartaric acid forming the bitartrate, or cream of tartar, which is gradually precipitated and removed. 1 g per litre of tartaric acid is removed by the addition of 1·5 g of neutral potassium tartrate per litre.

Concentration of must

Some attention now needs to be given to the use of concentrated grape juice for the enrichment of musts and wine. Sugar is now so expensive (say, £230 per ton) that an alternative source is worth exploring. But the sugar price would have to increase enormously before grapes became as cheap as cane or beet as a source of supply.

Grapes must now stand at, say, £500 per ton with 20 per cent of sugar. At this price, and valued only for its sugar content, the sugar is worth $5 \times 230 =$ £1,150 per ton. Alternatively, grapes, to be as cheap a source of sugar as cane or beet, would have to sell at

$$\frac{230}{5} = \text{£}46 \text{ per ton,}$$

a very low figure. However, the must adds not only sugar but other desirable constituents, including acids, all of which add quality to the wine.

Must is preferably concentrated by evaporation in a vacuum pan. It can, of course, be boiled down in an open vessel, but this tends to give it a 'cooked' flavour, so the low-temperature vacuum apparatus is preferable. In order to see how his must behaves on evaporation the wine grower can experiment with vacuum evaporation by adapting an ordinary domestic pressure cooker for the purpose. The exit valve (which usually is closed with a weighted stopper) is connected by means of flexible pressure tubing, incorporating a stopcock, to a laboratory filter pump. The pressure cooker is half filled with a known quantity of must and weighed. Gentle heat is applied and the pump started. After a time the stopcock is closed, the tubing disconnected and the cooker weighed again. The loss of weight will indicate the degree of concentration. The stopcock is needed to prevent water being sucked back into the warm must. When the liquid has lost about half its weight the process is stopped. The concentrate is allowed to cool, is poured off from any sediment and its gravity estimated in degrees Oe.

The quantity of must needed for enrichment can now be calculated, using the ordinary rule of proportion

$$x = Q\left(\frac{O_2 - O_1}{C - O_2}\right)$$

where: x = quantity of concentrated juice needed
Q = quantity of weak juice to be enriched
O_1 = degrees Oe of weak juice
O_2 = degrees Oe of desired mixture
C = degrees Oe of concentrated juice
obviously O_2 must be less than C.

EXAMPLE

Suppose we have 100 gallons of must of 62° Oe (potential alcohol 7·9 per cent) and wish to add some concentrated must having 125° Oe in order to get a must of 75° Oe, giving a wine with 10 per cent alcohol. How much of this concentrate will the 100 gallons need?

$$x = 100\left(\frac{75 - 62}{125 - 75}\right)$$

$$= 100\left(\frac{13}{50}\right)$$

$$= 26$$

26 gallons of the concentrate are mixed with the 100 gallons of the low-sugar must; the mixture will show 75° Oe and will ferment to give a 10-degree wine. As some or most of the acidity remains in the concentrate the mixture should be tested for acidity and steps taken to reduce this if needed.

Must can also be concentrated by means of partial freezing, the water forming ice, which is removed. It is not very satisfactory in practice as the ice removed carries with it a certain amount of sugar and other ingredients.

Freezing is, of course, the basis of the famous German *Eiswein.*

The frozen grapes are picked in November or later, usually during the night or about dawn, and carried straight to the press. Little or no pressure is applied and the liquid juice runs out, leaving the frozen water behind. Very high gravities are obtained in this way but very small quantities of must; for instance some frozen grapes in the Rheinpfalz were picked on the night of 16–17 January 1972 and the must had an Oeschler reading of 140°! The theoretical alcohol from such a must would be 20 degrees and no yeast could work at such a figure (most yeasts stop at about 15 degrees). The wine thus had a considerable amount of residual sugar. *Eiswein* is more a curiosity than a profitable undertaking. The risk in holding the grapes is considerable: birds, bees and rats attack them, for instance, and not much must is obtained in any case. As already mentioned, in 1974 Mr. and Mrs. Barrett, of Felsted, Essex, made some excellent *Eiswein* from their own grapes.

Picking should not start before the dew is off the grapes and should be done preferably during dry, warm weather. Pressing should start as soon as possible after picking. In this way damage, oxidation and rotting of the fruit are avoided.

Adding sugar to a must, of course, increases the volume and a useful figure in this connection is that 1 kg of sugar equals 0·6 of a litre (1 lb. of sugar replaces 9·6 fluid oz. of liquid, nearly half a pint).

Pressing

The grapes have been picked, the sugar content and acidity determined and the decision taken as to whether enrichment is needed or not. The next steps are pressing and fermentation. But before this is done another decision has to be made. Whether to remove the stalks before pressing the grapes or not. In general the stalks are not taken out for white wines, but a machine can be used which both crushes the fruit and removes stalks. In any case white grapes should be passed through rollers to crush the fruit thoroughly, because this facilitates the extraction of the rather viscid juice. The press itself does not always crush unbroken fruit, at any rate in the first pressing.

The stalks are usually taken out of red grapes intended for making red wines, because such wines will get enough tannin from the skins, but as whites do not stand on the skins for long they can do with the tannin from the stalks. The crushed grapes—

must—are taken directly from the crusher to the press, avoiding exposure to the air as far as is possible.

THE WINE-PRESS

Presses are of various kinds. There is the old-fashioned wine-press in which a threaded bar, made of wood at one time, but now of steel, turning in a fixed nut, is fitted into a stout frame. This bar, or tree, has a relatively coarse thread, deep cut to stand the pressure, and can be wound down, pushing the pressure plate into the cage holding the grapes. A somewhat simpler device is the static tree press: in this the tree is immobile in the centre of the cage and the pressure plate winds down on it as a stout bearing is turned. It has the advantage of dispensing with the heavy frame and the disadvantage that the steel tree passes through the grapes, which can thus pick up a certain amount of iron contamination. However, the worm of the tree is only in the upper part—from the top of the cage upwards—so that the lower part can be enamelled to prevent contact of the grapes with metal. In both static and moving tree presses the threaded part of the tree is fairly short, 2 feet, say, and wooden blocks are piled round it to transmit the pressure to the plate. When the plate has been forced down into the cage to the full extent of the thread the action is reversed, so that there is a gap between the pressure thrust bearing and the topmost block. More blocks are then put in and the pressure again applied.

The cages of such presses are usually made of oak battens, 2 in. × 1 in. section, set in a circle with gaps between them of about ⅛ in., smaller or larger according to the kinds of grapes to be treated. The cage is held together by a number of stout iron or stainless steel bands (three being usual), the bands being enamelled if of iron. The bands and the cage are usually made in two halves, with a device to join them securely, so that they can be taken apart for cleaning and removing the press cake (marc). Such presses, when small, are worked by hand through a series of levers and ratchets; larger ones are worked by hydraulic power (see page 130).

A cider-press can be used for grapes. This press is of the frame-work type, but with no cage; instead the crushed grapes are packed into press cloths, today usually of nylon fabric, and the filled cloths are built up on the bed of the press into a 'cheese' under the pressure plate. A framework of wooden battens is usually put between each filled cloth and the next to facilitate escape of the juice.

Modern wine-presses are worked electrically. A famous brand is the French Vaslin. Basically the Vaslin press is a horizontal stainless steel and fibreglass cage which slowly rotates while the two end plates approach each other. The juice is both squeezed out by pressure and sucked out by centrifugal force and gravity. Such presses are very efficient and can be electronically controlled to execute any desired pressing programme.

In any press the pressure should be applied slowly and steadily. When the fruit has been pressed almost to a solid cake the plate is wound back, the cage opened, the cake removed, broken up and then put back into the reclosed cage for a second pressing. A surprisingly large quantity of juice is usually obtained in this way. The cake can be cut a second time too and a third pressing made. *piquettes*, rather frowned upon by good vignerons and the EEC, are made by cutting up the cake once more, putting it back into the cage and adding water. The resulting juice has a slight must quality, needs a lot of sugar and makes only *piquette*, which can hardly be considered as wine.

Removing and cutting up the cake is a tiresome operation, and with the large hand- or hydraulically-operated batch presses a stainless steel spade is used to cut away a foot of marc all round the circumference of the cake and to throw this towards the centre of the press; then pressing is started again. The pressure on the upper part of the cake pushes the central part towards the circumference and much labour is saved in this way.

In the Vaslin presses this break-up of the cake is automatic. The two pressure plates within the rotating cage are connected by stainless steel chains just about horizontal and stretched when the two plates are farthest apart. As the plates approach each other the chains collapse into the fruit and eventually become embedded in the solid cake. When the plates are wound back the chains straighten out once more, breaking up the cake ready for a subsequent pressing. A great deal of hand work is thus saved.

Another efficient modern press system, much used in Germany, consists of a horizontal stainless steel cage in the centre of which is a strong rubber bag. The cylindrical cage is filled with the crushed grapes and the bag is slowly inflated with air, thus pressing the fruit against the cage. Because the rubber bag is flexible the pressure per square inch on it is equal over the whole surface, the pressure on the grapes thus being even throughout. After the first run the pressure is released and the cylinder rotated, which rearranges the grapes ready for another pressing.

Continuous action presses are also available, very suitable for the big vineyard. Basically they consist of an Archimedean screw of variable pitch turning in a horizontal stainless steel cage. The pulp is fed into the wide-pitched part of the screw and travels forward as the screw turns. The grapes thus move into the closer and closer pitched areas, and are subjected to steady and increasing pressure. The juice runs out from various orifices, corresponding to the first, second and third pressings of batch presses. Some modern presses of this nature work up a very high pressure at the end of the screw, delivering a dry marc more like sawdust than anything else. Not a drop of liquid is wasted; in fact it is possible that too much may be extracted from the fruit.

Needless to say, any press, particularly those with wooden parts, must be well cleaned before use.

Now for a practical example. Let us suppose we have 1,000 gallons of must showing 72° Oe and 14 g per litre of acid, and that we want a 75° Oe must with 12 g of acid. From the sugar table we find that 8 g per litre of sugar are needed to raise the gravity to the desired figure. 8 g per litre is the same as 8 g per 1,000 cc, or 0·8 per cent weight/volume, therefore 1,000 gallons of must need

$$\frac{0{\cdot}8 \times 10{,}000 \text{ lb.}}{100} = 80 \text{ lb. of sugar.}$$

2 g per litre of acidity will be removed by the addition of 2 × $\frac{2}{3}$ lb. per 100 gallon of calcium or potassium carbonate (see page 123), so 1,000 gallons will need 13$\frac{1}{3}$ lb. The quantities of these two materials are prepared and set aside.

Sulphuring

The vexed question of sulphuring now arises. Sulphur dioxide gas (SO_2), which forms sulphurous acid when dissolved in water, is a bleaching agent and preservative. Large amounts of SO_2 in the wine will keep it bright and clear for ever, destroy the bouquet and flavour and help its consumers to headaches and indigestion.

Consequently it should be the aim of the vigneron producing a quality wine to use the minimum quantity of this substance.

SO_2 gas (and a little SO_3) is formed when sulphur is burnt, hence the fumes from burning sulphur are valuable agents for cleaning barrels and apparatus. For instance, an old-fashioned wooden wine-press can be washed with water and then purified by being covered with a hessian or canvas sheet, underneath which a ladle of burning sulphur is pushed so that the fumes penetrate all round the cage and pressure plate. Barrels may also be purified by lowering burning sulphur wicks into them and closing the bung. The wick is best put into a small tin with some holes in the sides, to prevent any burning sulphur bubbling over into the barrel. Apparatus can also be cleansed and purified by being washed with SO_2 solution obtained either by dissolving the gas (it can be obtained compressed into cylinders) in water or by dissolving sodium or potassium metabisulphite in water, when SO_2 is released. A suitable washing solution is made as follows: make up a 10 per cent stock solution by dissolving 1 lb. of metabisulphite crystals in warm water and making up to 1 gallon volume with water. For use 1 pint of stock solution is added to 9 pints of water.

Sulphuring the must prevents oxidation, destroys harmful bacteria and ferments, and delays the start of fermentation a little, which last can be a disadvantage if grapes are coming into the cellar in large amounts, as the vats are occupied for longer periods than in the case of unsulphured juice. A light sulphuring is certainly advisable in commercial practice, but it must not be overdone or the sulphur will carry through to the wine and give it that horrid sulphur taste all too common in cheap white wines.

As to the amount to use—Chancrin[20] suggests 10 to 25 g per hectolitre for whites, while Nègre and Françot[75] opt for 25 to 30 g. The EEC regulations lay down maximum SO_2 limits on wine offered for sale and the British grower would be well advised to regard these figures as maxima and to try to get below them if possible. They are, for white wines:

With 5 g per litre residual sugar or more—300 mg per litre SO_2.
With less than 5 g per litre residual sugar—250 mg per litre SO_2.

These are high levels, respectively 30 g and 25 g per hectolitre. S. F. Hallgarten thinks these figures are so large because a new method of analysis has been introduced and that in reality they may be below the old legal levels.[45] It should be noted that the permitted level in Britain is not more than 450 mg per litre in the wine. The British drinker of cheap white wine gets a bigger dose of SO_2 than his EEC counterpart.

As batches of must come from the press they are immediately sulphured, using either a strong solution of potassium metabisulphite or gas from a cylinder. Since sulphur is lost in processing the dosage can be calculated for, say, the 25 g per hectolitre rate (equal to 2·5 lb. per 1,000 gallons). The amount of gas drawn from a SO_2 cylinder is obtained by weighing the cylinder at the start of and during treatment. The cylinders must be treated with great respect as SO_2 is poisonous.

Cleansing the must

The next process is to clean the must of suspended matter, such as soil, spray residues, broken cells from skin, stems and pips. The old-fashioned method (*débourbage*) was to allow the sulphured must to stand twenty-four hours in a cool place. Most of the solid particles fell to the bottom of the vessel and the clear liquid could then be drawn off. A certain amount of must was wasted, namely that held in the precipitate.

A more modern method is to use a centrifuge. Acting on the cream-separator principle the centrifuge turns at about 9,000 revolutions per minute. The particles go to the wall of the chamber and the cleaned must is delivered from the centre; there is thus no waste. The action is continuous. If deacidification is to take place it is best done just before the cleansing operation: in this way the precipitated acid is removed with the dirt, etc.

Fermentation

The sugar is now added to the must and, when it is dissolved, the deacidifying agent, unless previously used.

Sulphuring the must destroys bacteria and some wild yeasts and delays fermentation by the wine yeast, a single-cell plant—*Saccharomyces cerevisiae* var. *ellipsoidus*. In order to avoid delay and get the must active it is usual to employ a 'starter', that is, a small quantity of actively fermenting must, which will set the whole batch going and give a slow and cool fermentation. It also allows the bulk to be seeded with one's own pet yeast, though I doubt whether this makes much difference in the end. The use of special yeast strains is the subject of much myth and mystery. Hallgarten takes the view that 'foreign' yeasts make very little difference; for instance, a Champagne yeast used on German must does not produce a wine with Champagne characteristics.[45] Logic would also suggest that, though the must be seeded with a foreign yeast, the local ones, those on wild fruits and floating around in the air, being adapted to the neighbourhood—the survivors in the Darwinian struggle for existence—will take over in the end; that is, unless the must is completely sterilized by vast amounts of SO_2, or by pasteurization, and then fermented under strictly aseptic conditions, almost impossible to do at any reasonable cost. However, it does no harm, except to the purse, to use a pure strain of a foreign yeast in the starter and it can at any rate get the wine going in the right direction at the commencement of the process. In this case, say, a gallon of must is heated to about 80°C, allowed to cool and is seeded with the desired yeast, foreign or local, put into a clean glass vessel which it should not fill (a carboy, for instance), and plugged with cotton wool. It should be kept warm and will be fermenting briskly when the main batch of must is ready. The starter is then poured into the bulk, stirred and the whole set aside for fermentation.

The yeast plant in order to grow must use oxygen (among other things) and it can get this from the air or from sugar. Our aim, of course, is to make the yeast draw on the sugar for its oxygen. In the early stages of fermentation, for instance with the starter, the must is exposed to air in order to get a rapid multiplication of yeast. After that, air is excluded, either by carrying out the process in full vessels or by allowing a layer of CO_2 gas to accumulate over the surface of the liquid. The yeast then draws on the sugar for its oxygen, converting it to alcohol and CO_2 gas in roughly equal parts, by weight.

	$C_6H_{12}O_6$	=	$2C_2H_5OH$	+	$2CO_2$
	sugar	=	alcohol	+	carbon dioxide gas
Molecular Weights	180	=	92	+	88
Per cent	100	=	51	+	49

Another decision now has to be made. This is on the filling of the fermentation vessels. They can be either about $\frac{7}{8}$ full or filled to the bung (in the case of barrels), when the escaping gas carries out a lot of froth, pips, dirt, yeast and other extraneous matter that may have escaped the cleansing process; but it also carries out a little half-formed wine. Some growers think the latter method (known as *guillage* in France) is particularly suited to making white wine from white grapes, that it gives a better colour in the resulting wine and is worth the small loss that occurs. I must say I rather like it myself. If the *guillage* process is being used it must be remembered that as the gas escapes the level of the liquid is reduced, hence topping up is required from time to time. Obviously it is best to use must for this process, but see page 136 for a cautionary lesson. Part of the sugar, or unfermented must, can be used for this purpose. If the alternative system is used, the spent yeast settles into the lees after fermentation has died down. The wine tends then to have a slightly more pronounced 'yeasty' taste, which is not necessarily disadvantageous.

In the *guillage* system, after the violent fermentation has died down, a fermentation lock of some sort is put over the orifice in order to prevent the entry of air, and the slow final process is allowed to terminate.

Heat is produced during fermentation, but it is unlikely to be excessive under English conditions. The winery should be well-ventilated, because the accumulation of CO_2 gas can be dangerous to the workers. The gas is not poisonous in itself, but it is heavy and can replace oxygen in a cellar environment, leading to suffocation.

Racking

After fermentation has ceased the wine must be kept in full containers and allowed to settle. The yeast, in its spore form, gradually settles to the bottom and in due course the wine is 'racked'; that is, the clear, or more or less clear, wine is drawn off from the lees. The wine may be left on the lees for varying lengths of time, according to the taste and experience of the wine maker. In general, for white wines, the first racking should be in December. If the wine after this racking is not clear, consideration must be given to 'fining', that is clarification by certain additions or methods.

If the wine is reasonably clear a second racking, say in February, may clear it sufficiently without any further processing, particularly if it has been in a cold place during the winter. Most wines do not require fining. A critical examination by appearance and taste enables the vigneron to judge whether the additional process is necessary. On big estates a chemical analysis is made after the first racking to see if any further treatments are necessary.

Fining

The object of fining the wine is to precipitate any solid matter suspended in it. Obviously this can be done by filtering it, but often at this stage the solids in the wine are in a very finely divided state, almost colloidal, so that filtering is difficult because the filters become clogged very quickly.

The precipitation can be secured by mechanical means or chemical ones.

MECHANICAL FINING

1. *Centrifuge*

If a centrifuge is available the wine can be put through this, but it may not remove the finest particles.

2. *Sterilizing filter*

In spite of the difficulties, sterilizing filters are now available, but they are generally used on the bottling line when there is not much, if any, sediment to be removed. These filters remove bacteria, yeast spores, and other solid materials. It is a common practice first to centrifuge the wine to remove the coarser particles and then to filter it.

3. *Cold*

Cold has the effect of clarifying wine and the British wine-grower can thus take advantage of our climate by having some storage vats or barrels in an outside shed or barn. A considerable quantity of potassium bitartrate is thrown down in cold weather. The wine should be brought down to about −3°C.

4. *Clays and other additives acting mechanically*

Kaolin, Bentonite and even sand stirred into the wine fall to the bottom and take the wine solids with them. If sand is used it must be a silica one. A shell sand, containing calcium carbonate, would reduce the acidity of the wine. Torn-up paper has also been used for fining.

CHEMICAL FINING

This depends on fining agent combining with tannin in the wine and producing an insoluble substance which then falls to the bottom of the vat. Among the substances used are: white of egg (much used for white wines, 2 egg-whites with a pinch of salt (20 g) per hectolitre (22 gallons) of wine); blood albumen; gelatine; isinglass; and a number of proprietary fining agents. The process removes some tannin from the wine; a white wine may already be a little deficient in this substance, in which case some tannin must be added at the same time.

After fining the wine is allowed to settle for two weeks and then racked off into cleaned and sulphured containers. It is now ready for further treatment, which can be sweetening, storage (ageing in cask, for instance), or bottling.

Sweetening wine

Some consumers like a sweet or sweetish wine, and if the grower is catering for this market it is easiest to add the sweetening at the last moment, just before bottling. The process is known as *dosage*.

Under the EEC regulations sugar may not be used for this purpose, but grape must, natural or concentrated, may be. Obviously the sweetner must not contain any yeast or the fermentation process might start again in the bottle and blow out the corks. Under German law the amount of sweetening which may be used also is restricted and depends on the amount of alcohol in the wine. The remaining sugar in the wine must not be more than $\frac{1}{3}$ the weight of alcohol there. Thus a wine with 70 g per litre of alcohol (8·82 degrees) may not have more than 23·5 g of sugar per litre. A stronger wine, say 88 g per litre alcohol (11·1 degrees), is limited to 30 g per litre of unfermented sugar.

To prevent this sugar fermenting the wine may be pasteurized as it flows to the bottling machines, but this tends to produce a 'cooked' taste in the product and it is more usual to pass it through a sterilizing filter and to use 'sterilized' bottles and corks. Neither is really sterile, but they have been heated enough to kill any yeasts present. Some very careful German wineries actually use 'sterile' bottling rooms, in which elaborate precautions are taken to filter the air going into the room, reminding one somewhat of an operating theatre. Of course, neither such a bottling room, nor the operating theatre, is really sterile, but the risks of infection are considerably reduced by such plants.

Wine can also be pasteurized after bottling and, of course, before labelling. Low alcohol wines are heated to 65 °C and ordinary wines to 60°C.

Bottling

Most wineries of any size have an automatic bottling, corking, labelling, capsuling machine and corks are softened and sterilized by boiling for a short while before use.

Corks are now so expensive that some thought needs to be given to the use of substitutes, plastics and crown cork sealings. The Champagne cork is the height of nonsense. Firstly, cork of the requisite size is so scarce that most Champagne corks today are made of cork slices stuck together, and secondly, the whole object of the cork is to keep the gas in the bottle, a thing which would be done much more efficiently by a crown-cork closure. In fact crown corks are usually used for bottles in the maturation process, up to the *dégorgement*.

Most bottles today are hermetically sealed, that is, the cork is capped with metal or plastic, allowing no air to reach it: consequently the bottle might just as well be closed with an air tight, and much cheaper, plastic seal. However, there is a view among customers that a plastic or crown cork seal means a cheap and inferior wine and that a cork somehow allows the wine to improve in bottle. There may be a modicum of truth in this. The cork does allow a tiny exchange of gases: possibly some alcohol, CO_2 and esters come out and a little air goes in. The quantities involved must be very small and the oxygen going in can only lead to oxidation in the wine, a thing the vigneron is at pains to prevent throughout the whole vinification operation. Against the view that the cork is just there for snob appeal, is the fact that many fine German wines are now exported with a lead capsule having three small holes punched in it, allowing air to reach the cork.

Some interesting tests were made in this sphere in Germany recently by the Geisenheim Viticultural College.[45] A 1970 Silvaner wine, bottled in February 1971, was used, some being sealed with corks and some with plastic (screwcap closures). Blind tastings with four groups of tasters gave the following results:

'three groups unanimously gave the preference to the screwcap wine.

'fourth group divided: one preferred the screwcap wine, four were for the corked and one thought both wines equal.'

Can the snob value of the cork be defeated and things made easier for the wine grower and consumer? It is much to be hoped that it can.

Summary: To make a 10-degree wine

The grapes should be picked on a sunny day in early October and immediately pressed. A sample of the juice is examined with either a refractometer or a hydrometer and the table (page 158) consulted to see how much more sugar, if any, is needed to result in a wine of 10 degrees (10 per cent by volume) of alcohol.

The must is also examined for acidity by titration with an alkaline solution (see page 123). If it is over 7 g of tartaric acid per litre steps should be taken to reduce it, by adding carefully calculated amounts of chalk, 'Acidex' or potassium monotartrate (see page 124).

The must is then treated with small amounts of sulphur dioxide (or Campden tablets, one per gallon). After this deacidification, if needed, takes place.

The must is now cleansed, either in a centrifuge or by being allowed to stand for twenty-four hours. In the latter case much solid matter is deposited and the relatively clear must is then drawn off. The sugar, if needed, is now added.

The must is now ready for fermentation. A 'starter' is usually used. This is a small quantity of must to which a desired strain of yeast has been added, the whole being kept in a warm place. When fermenting briskly the starter is added to the bulk of the must and sets the whole lot going.

Fermentation may be done in $\frac{7}{8}$-filled vessels or in completely full, necked containers, or barrels on their sides (the *guillage* process). In the latter case spent yeast, odd pips and bits of pulp and skin are thrown out with the bubbling. Obviously the bulk is reduced in this way and the containers or barrels must be topped up from time to time with fermenting must from another vessel.

When the brisk bubbling has died down, a fermentation lock is fitted to the vessel to prevent the access of air and bacteria. Such a lock is usually an arrangement of tubes and bulbs containing metabisulphite (Campden) solution, but a plug of cotton wool can be used instead.

When fermentation has stopped completely the wine is transferred to storage vessels (such as barrels) which must be kept full.

If these vessels can be kept in a cold place clarification will be assisted.

During storage the wine clears and it is then drawn off from the lees, usually in December. Another racking may be given in February. If, by then, the wine is not bright, it must be fined by the addition of 'finings' (see page 135).

The wine is now ready for bottling, which is best done in cold weather. If a sweet wine is required some unfermented must, or must concentrate, may be added just before bottling, but care must be taken that no yeast gets into the bottle, or fermentation will start again and blow out the corks. This can be prevented by passing the wine through a sterilizing filter, or by heating the wine, or filled bottles, to 60°C (pasteurization).

9
Some factual and commercial considerations

The yield of grapes from an English vineyard in any but a disaster season will range from one to five tons per acre, and unless the grower gets two tons or more he is unlikely to make money from the enterprise. A bottle of wine needs from 2·3 to 2·5 lb. of grapes, so the above figures represent 896 to 4,870 bottles per acre.

At the English Vineyards Association conference at Kenilworth in 1975 some interesting figures were put forward by Messrs C. D. Walker and R. S. Don, both authors—particularly the former—stressing the general nature of the entries.[26, 114] Obviously, as Mr. Walker pointed out, each grower or would-be vigneron must make his own estimates as there are so many variables. For instance, Mr. Walker's figure for bottles per ton of grapes is 975, or 2·297 lb. per bottle, which is also the Merrydown figure. In Mr. Walker's example a five-acre vineyard cost £4,184 per acre up to the end of the third year, when the first return might be expected. The figures did not include land but did contain everything else (plants, posts, wire, working, picking, processing, duty and interest at 15 per cent). The cost per bottle, duty paid, from Mr. Don, at the 1-ton rate, was £1·08 or about £968 per acre, which is not a large sum to set against the accumulated establishment cost of £4,184, but which if sold at £1·20 per bottle would give a 10 per cent profit on turnover.[26] In subsequent years the crop would increase and the variable costs (running costs) would remain much the same, so that by the fourth or fifth crop (the seventh or eighth year after planting) all the accumulated costs should have been liquidated and the vineyard be making a clear profit. For instance, with a cost for growing of £900 per acre and for processing 4 tons of fruit of £2,216 (including duty) the cost per bottle is said by Mr. Don to be 49 pence, though I do not quite follow his calculation. But obviously the greater the crop the less the cost per bottle, because the growing costs are the same for a 1- or a 4-ton crop. Picking and wine-making costs are obviously more for the bigger crop. It should be noted that Mr. Don's figures were for the 1974 excise duties which have been much increased since then.

The above figures may not seem very encouraging to the would-be wine grower but there is no doubt that considerable economies can be made all along the line. It must be realized that any writer putting up figures of this sort must have them on the high side in order to be safe. Messrs Walker's and Don's figures are a most valuable guide. Every would-be grower should carefully peruse the original papers, enter his own figures and see how the sum comes out; in fact in the duplicated edition of his Paper Mr. Walker left a blank column for this very purpose.

The economies that can be made are in buying plants, avoiding unnecessary cultivations or sprayings in the vineyard and in having sufficient acreage to make it worth while establishing one's own winery. Better wine will be made by not having to transport grapes long distances, transport and other costs will be avoided and the grower will have the satisfaction of having the whole process under his own control. Paying for the services of an expert wine maker is well worth while in the early stages of production. Of course, at first it is a great advantage to have one's precious fruit put into the hands of an experienced wine-maker and at all times it will save the grower a great deal of work. But it is opting out of half of the joy and business of the matter and will be expensive: fresh grapes too make better wine.

The price of grapes in continental Europe is now remarkably high. For instance, champagne is now in much demand: the production per acre and the area planted to vines are both rigidly controlled and cannot be increased. The pattern in the Champagne is that much of the fruit is produced by a large number of small producers selling grapes to the famous Champagne houses (which have their own vineyards as well). The 1975 price for top quality Champagne grapes was fixed at 6·38 fr. per kilo or 32 pence per lb., say £720 per ton, and this was 3 francs per kilo less than in 1974. The price, however, would be considerably lower in other areas, say £500 per ton. Nevertheless that price means that the grapes alone in a bottle of champagne have cost over 60 pence. The remarkable thing about that wine is not its expensiveness but its cheapness in view of the materials and labour needed to make it.

Customs and Excise

The British vigneron is particularly hard hit by taxes on his product, and an ingenious device has been incorporated into the regulations whereby a tax is levied on a tax. Value Added Tax at 8 per cent is payable on the sale price of wine, which already includes a considerable element of tax. The present (Sept. 1977) duties on table wines, home-produced or imported, are £3·25 per gallon (£0·5417 per bottle) plus 8 per cent Value Added Tax, which, if the wine is sold at, say, £1·75 per bottle, is another 14 pence. Thus the tax on a bottle costing the consumer £1·75 is 68·2 pence, or 39 per cent. As Sir Guy Salisbury-Jones has pointed out, this is equivalent to taxing a promising new industry at a rate of more than £1,000 per acre, assuming a modest crop of just over 2 tons per acre.

Wine is more highly taxed than beer in terms of alcohol content, as may be noted in the table below:

Cider, beer, wine and spirit duties, 1977

Product	*Duty per gallon* £	*Equivalent in pence per fluid oz. of alcohol*
Cider	0·242	2·75
Beer (5%)	0·548	6·85
Table wine 15%	3·25	13·5
10%	do	20·3
8%	do	25·4
'Made wine' 15%	3·16	13·2
10%	2·11	13·2
Spirits	£27·165 per proof gallon*	29·79

Source: Commissioners of Customs and Excise. Notices 957, 959. Mar. 1977.

* Proof spirit is 57% alcohol, consequently a gallon of proof spirit contains 91·2 fluid oz. of alcohol.

It is a curious anomaly that the promising new industry should be so heavily penalized in the present new regulations, which must give great comfort to the established manufacturers of 'made-wines', brewers and even distillers. The duty on the alcohol in a table wine of 15° (15 per cent) is about double that on the alcohol in beer and as the same duty per gallon is paid on all table wines up to 15°, the duty on the alcohol in a light wine is three and a half times that on the alcohol in beer. In the case of an 8° wine the duty per ounce of alcohol is approaching that of spirits. That is not all: 'made-wines' which in commerce are mostly those manufactured in Britain from imported grape concentrates, pay a lower rate of duty and have an additional favourable classification of 'under 10 per cent alcohol'. The figures are:

	Duty per liquid gallon £	*Per bottle* £
Wine, not exceeding 15%	3·25	0·542
Wine, 15–18%	3·75	0·625
'Made-wine', not exceeding 10%	2·11	0·352
10–15%	3·16	0·527
15–18%	3·475	0·579

It is difficult to see the reason behind the favouring of the lower quality 'made-wines' at the expense of the true wines, whether of British or EEC origin. Since the grapes for 'made-wine' are not grown in Britain, less British labour is employed in their production. Moreover the imported pulp has to be paid for in scarce foreign currency. Perhaps good wines are thought to be luxuries for which the consumer should be forced to pay dearly if he is not content with the common man's beer, but that if he must drink wine, then he should consume the products of existing manufacturers of 'made-wines' and not disturb the *status quo*. As already noted, it is much easier for the Customs and Excise to check the books of a few large manufacturers than to visit several hundred wine growers in Britain for the same purpose—possibly a powerful reason for the preference shown to made-wines.

The business done in such wines is considerable. For instance, the *Business Monitor*, PQ 239.2 (HMSO, 1975) reports that for the year ending 21st September 1975 the production from nine manufacturers was 15·7 million gallons, worth £49·3 million. These figures include fortified wines, such as the 'port-type' and 'sherry-type' products, but much of the total was for table wines, such as those made from British-grown grapes. The substantial differential these manufacturers now enjoy over the British vigneron must be very welcome to the former.

Does the British vigneron have any remedy? Three occur to me. First, to make appeals to the Office of Fair Trading and the European Economic Commission. Second, to ship the must to, say, Calais or Le Havre, then reimport it and hope to enjoy the same preferential rate given to manufacturers employing foreign (and mostly non-Common Market) musts. This would probably not be allowed: the Briton would still be penalized for trying to disrupt the established order. A third method, absurd as it may sound, would be to turn one's wine into made-wine by concentrating the must in a low temperature vacuum pan and then reconstituting the must by putting back the same amount of water just taken out. It would be a waste of fuel, but by qualifying under the law as a made-wine manufacturer, paying the favourable rate of duty, it would be worth the grower's while: it would also have the advantage of being in accordance with much of today's economic idiocy.

In the present economic climate, and with our membership of the European Economic Community, it would be idle to expect that future legislation would show any preference towards home-produced wines (which existed at one time), but growers could urge that at least they be not taxed at higher rates than made-wines (which would only be a reduction of about 3 per cent) and that a category for 10° wine be made for our own (and EEC) true wines, at the same rate as for made-wines (namely £2·11 per gallon), which would be a reduction of 35 per cent in tax. One supposes it would be too much to ask for a reduction to a figure resembling the excise on beer. The only other hopeful aspect is that the EEC might order Britain to reduce duties on wine to nil, or at least to the current small figures within the rest of the community. The duties within the EEC at December 1975 are given in the Table. How fortunate are Germany and Italy!

Two minor, but none the less welcome, changes in Britain are that the Customs and Excise have stopped calling wines by the strange name of 'sweets' and that they now measure alcoholic

Table wines, 12° or less	*Units of Account per litre*	*Approx. equivalent £ per gallon*
Belgium	0·130	0·339
Denmark	0·773	2·017
France	0·017	0·044
Germany	nil	nil
Ireland	0·666	1·738
Italy	nil	nil
Luxembourg	0·130	0·339
Netherlands	0·130	0·339
United Kingdom	1·006	3·25 plus 8% V.A.T. = £3·51

strength in terms of per cent alcohol by volume, as well as in degrees proof. Proof spirit contains 57 per cent by volume of alcohol: a wine of 26·2 degrees proof contains

$$\frac{26{\cdot}2 \times 57}{1000} = 15 \text{ per cent of alcohol by volume.}$$

The strength of German wines is usually reckoned in percentage of alcohol by weight: since alcohol is lighter than water the by-weight figure is lower than the by-volume figure for the same wine. The by-volume figure multiplied by 0·802 at 8 per cent (by-volume) up to by 0·809 for 15 per cent (by-volume) gives the by-weight figures, i.e. 6·5 per cent and 12 per cent. The variation in the factor is caused by the fact that if you mix equal volumes of pure alcohol and pure water, say a pint of each, you will not get 2 pints of liquid, but something less. The respective molecules fit in beside each other somehow or other and warm up too. Many an unfortunate cellarman has unjustly been accused of misappropriation because of ignorance of this unfortunate physical fact.

The British vigneron today is in a much better position as regards paying duty than he was twenty years ago. Then he had to have a double-locked bonded store, the Customs Officer holding one key and he himself the other, so that access to it could only be obtained in the presence of both parties. Today he must obtain a licence to sell wine in bulk (per case or more) costing £5·25 and keep books on his transactions for inspection by the Customs Officer. He is then allowed an open bond and only pays duty as the wine is sold. The home wine maker does not have to have a licence or pay duty provided he does not sell any wine, or barter with it. Until quite recently the professional vigneron had to pay excise on any wine he used in his own home but from 1st January 1976; 'Wine produced from grapes grown in the United Kingdom may be sent out from a winery without payment of duty for the domestic consumption of the grower in such quantity as the Commissioners may on application from him allow.' (Regulation 20(2) of the Wine and Made-Wine Regulations, 1975.)

This amount is 120 gallons plus 10 per cent of his annual production over 120 gallons, subject to a maximum of 240 gallons in total. This is a generous concession, the maximum being equal to just under four bottles a day. While it will be welcomed byBritish vignerons it is but a sop (almost literally!) to them: what they really need is to be treated on the same terms as the manufacturers of 'made-wines'. For instance, the new concession will allow, say, 400 growers 150 gallons each, a total of 60,000 gallons, a quantity which otherwise would have paid £195,000 in duty, and apparently a considerable gift from the Customs and Excise to the English vignerons. None of this concession wine may be sold but, presumably, part of it can be used in promotional work in the home, stimulating the demand for English wines. The English vineyard produces, say, 200,000 gallons per annum and pays (at £3·25 per gallon) £650,000 in duty. Were this English production taxed at the made-wine rate it would only pay £422,000 (if 10 per cent alcohol or under), which is not only a saving of £228,000 (£33,000 more than the concession gift) but would also allow the wine to be sold. British growers would far rather have this equality of taxation than a concessìon that can only be enjoyed by the process of drinking their own profits, pleasant as the exercise may be.

Distillation

Home distilling is not allowed whether the product is for sale or not, unless licences (costing £15·50) has been obtained and the regulations obeyed. The punishments for infringements, one is led to believe, are frightful—I am not quite sure what, but 'something lingering, with boiling oil in it' as Gilbert's Mikado said to his would-be son-in-law.

Mr. Barrington Brock, who has already been mentioned, found that English wines made good brandy, and Mr. Anton Massel, the well-known oenologist at Ockley, Surrey, is of the same opinion.[66]

The conditions under which a would-be distiller would have to work are governed by the Customs and Excise Act, 1952, and the Spirits Regulations, 1952. One of the most important of these is Section 93 (4) which lays down that a licensed still must not be less than 400 gallons capacity. Obviously the would-be brandy maker must work on a fairly big scale and Customs and Excise prefer dealing with large concerns. One filling of the minimum size still with a 12° wine would theoretically boil off enough spirit to make 48 gallons of absolute alcohol. In a pot still (the best for brandy making) not all the alcohol is recovered. In practice let us assume that 45 gallons of absolute alcohol is obtained, which is equal to 79 gallons of proof spirit (57·06 per cent alcohol by volume). The duty is now £27·165 per proof gallon so that the sum due on each filling would be £2,146. Brandy is usually sold at 70° proof, thus the 79 gallons of proof spirit would make 113 of 70° proof, or 678 bottles.

$$\frac{2146}{678} = £3{\cdot}16 \text{ per bottle duty.}$$

For how much could a bottle of English brandy be sold, duty paid, off the winery? If, say, £6 this would attract another 48p. Value Added Tax, leaving the grower with £2·65 (6 − 3·16 − 0·48 = 2·36) to cover growing the grapes and manufacturing. The bottle of 70° brandy will have been derived from 8 to 10 lb. of grapes. If half the ex-bond returns on a bottle (say £1·18) is allocated to the growing of the 9 lb. of grapes and 3 tons (6,720 lb.) per acre are grown this is equivalent to

$$\frac{1{\cdot}18 \times 6720}{9} = £881 \text{ per acre.}$$

Though not a large sum, it is a possible figure. On the other hand perhaps £6 per bottle for English brandy is rather optimistic.

Publicity

There is a considerable degree of local interest in a vineyard. At first the enterprise is regarded as a foolhardy venture and the watchers gleefully await its failure. After some judicious local tasting and talkings, in which the Women's Institutes must not be forgotten, a great deal of regional patriotism will be generated for your product.

The grower must also decide whether he is going to accept 'tours round the vineyard' from coach parties and tourists generally. This is quite another business and must be entered into as such. If it is undertaken the vigneron must smilingly answer the same daft questions day after day. To the individual who wishes to know if he can come and tread the grapes with his feet the proprietor can hand a hoe, saying 'No, but it would be fine if you would weed that row for me'. He will not get many takers. A good point is that visitors usually buy some wine.

Mr. R. S. Don's advice on these points is very sound and should be read by all interested in the subject.[26]

10

The amateur grower and wine maker

The amateur grape grower and wine maker does not have a lot of money at stake and thus can experiment in all sorts of directions that would be far too risky for the professional. For instance, he can make, if he so wishes, very 'natural' wines, similar to those of fifty years ago. Today a commercial wine must be bright and crystal clear and stay so to the end of time. This was not the case in the 1920s. German wines were drunk from long-stemmed glasses having decorated bowls, usually tinted yellow, brown or green. The glass concealed the fact that the wine was not 'bright' and it was none the worse for the little cloudiness it had.

An amateur's wine, that on ageing in bottle throws a deposit of fine, white crystals, looking like sugar, is not to be condemned. The crystals are the double tartrate of potash and show the wine's purity and acquisition of desirable qualities as it matures. Home-made wine for home consumption attracts no tax and is consequently cheap; nor is one forced to drink every drop in the bottle—a few dregs can be left, or the wine decanted. Moreover the amateur does not have to stuff his wine full of SO_2, fine it endlessly, sterile-bottle and fuss over it as if it were a delicate child. Leaving it alone, he may lose a few bottles here and there, but he will have a natural wine most of the time and be gaining valuable experience in both growing and wine making. It is a living wine, unsterilized.

He can experiment with old and new varieties, make sparkling wine by the *méthode champagnoise*, country wines too if he feels like it, and he may keep his first, second and third runs from the press separate and compare them, because he can ferment small batches. He can also keep any promising vintages to let them mature, a thing the commercial grower seldom can afford to do with the high interest rates prevailing today. In short he is involved in a subject of the greatest fascination. But I repeat that there are two things he must *not* do; sell any of his wine or exchange it for goods, or distil any of it. If he does, he risks great trouble from the Customs and Excise, which has eyes everywhere. Remember that illegal stills are always discovered. Two things betray them. First the smell, both of the operation itself and that of the dunnage (the residue left in the still after distillation), which has to be thrown away. Both are strong and characteristic; the smell around cognac can be quite ghastly. Secondly, if you illegally distil and get a pleasant brandy, your pride in it will go before a fall. At your vintage celebration party, after a glass or two, you will say to a friend: 'What do you think of this, old chap? Made it myself.' Standing behind you will be your friend's friend, the person he brought to the party, who is the local Surveyor of Customs, or one of their minions . . . 'Alas! What ill-timed bonhomie.'

Nor may you concentrate the alcohol in wine by freezing it. In the first place it is not a very satisfactory process and secondly, unless licensed, it is apparently illegal too.

A thing the amateur should do is to keep a diary and note-book: much valuable information will be collected.

The amateur, if anything of a handyman, can make much of his own equipment. An old-fashioned mangle with wooden rollers is easily converted to a grape crusher and the press itself presents no insuperable difficulties. The home-made article may not extract as much juice as a modern pneumatic or Vaslin press, but it will give a good must without any need for measures against excess tannin, broken pip and stalk flavours, which may occur if heavy pressure is used.

The making of a press has been described in many books, including one of mine, so I only give a few hints on the subject here.[47, 78] The cage is the important feature. Let us make a calculation on this. The cage is held by, say, three iron bands and is to be composed of oak battens, say 2 in. × 1 in. × 24 in. long. Let us assume we want it about 2 ft in diameter. To the top of the battens it will then hold $\pi r^2 \times 24$ cubic inches, or $3{\cdot}14159 \times 12 \times 12 \times 24 = 10{,}857$ cu. in. A cubic inch equals 0·56767 of a fluid ounce and there are 160 fluid ounces to the gallon. Therefore the projected press's capacity is

$$\frac{10{,}857}{0{\cdot}568 \times 160} = 119 \text{ gallons}$$

or, say, 10½ cwt of grapes. The internal circumference of the cage with a diameter of 2 ft is $d\pi$ or $24 \times 3{\cdot}14159 = 75{\cdot}4$ in. The battens are 2 in. broad and a gap of ⅛ in. must be left between them to let the juice out, therefore

$$\frac{75{\cdot}4}{2{\cdot}125} = 35{\cdot}48, \text{ equals the number of battens required,}$$

that is 35 battens at 2·125 in. broad plus a gap of 1·025 in. A special batten could be cut to fill this, but it would be neater either to reduce or increase the diameter a little by using one less or one more batten. An even number of battens is desirable in case the cage is being made in two halves (a cage in two halves makes cleaning much easier), when obviously the same number of battens on each half would be required. Suppose we increase the circle to take one more batten and its gap—2·125 in. We already have 1·025 so we need $2{\cdot}125 - 1{\cdot}025 = 1{\cdot}1$ in. more. This addition increases the circumference to 76·5 in. and the diameter

$$\text{becomes 24.4 in. } \left(\pi = \frac{76{\cdot}5}{3{\cdot}14159} = 24{\cdot}4\right)$$

The rings must be forged by a blacksmith. The smith will be more interested in the length of the circumference than in the diameter. Three strips are cut 76·5 in. long plus any necessary overlap (if any) for joining. One strip is marked for drilling with a punch at intervals of 2·125 in., put on top of the other two, and all are drilled together. A saw cut is made on the three bands so that they can be aligned in assembly. The drilled holes are then countersunk on one side and forged to a circle with the countersinking outwards. If the grower wishes, they can then be painted with several coats of a plastic base enamel, or they can be warmed and dressed with paraffin wax. Even if left untouched, the juice running from the press cage hardly touches the bands, so that the wine will not be contaminated, but, of course, the hoops will rust and have to be cleaned. The cage is assembled with suitable wood screws. While stainless steel screws are undoubtedly best, ordinary wood screws can be used, being rolled in the enamel or warmed wax before use. Stainless steel, fibreglass and plastics are now used for apparatus (steel for presses and press bands and containers, and fibreglass and plastics for the latter as well). Although stainless steel presses are mostly employed by professional vignerons there are one or two small models available, such as the Vaslin (see page 129).

The cage is the only difficult part: making the rest is easy and will be found, as I mentioned before, in many books. A pneumatic car-jack is likely to exert more pressure than most screw-thread systems. It is convenient to think of the power of the press in terms of the 'magnification' effect. I have a small French press in which the tree has 10 threads in a distance of 12 cm, or 1·2 cm per thread (the thread is square cut and coarse). When the operating lever has made a circle the plate will have descended (or ascended) 1·2 cm. The lever is 130 cm long: therefore, from the formula $d\pi$ for the circumference of a circle, the end of the lever will have travelled $260 \times 3{\cdot}14159 = 817$ cm in making one revolution. The 'magnification' then is

$$\frac{817}{1{\cdot}2} = 680.$$

1 cm descent of the plate is secured by 680 cm travel of the end of the operating lever. My press has a ratchet arrangement, slightly increasing the magnification and enabling the press to be worked from one position; the operator does not have to walk round and round the cage. The closer the threads and the longer the lever, the greater the magnification, but too close a thread will wear quickly and strip under pressure.

A press must be securely fixed to the floor, or to a platform on which the worker is standing.

The pressure per square inch exerted on the grapes depends on the amount of push on the end of the operating lever. Let us suppose it is 112 lb. (8 stone). Then 112×680 lb. is the downward thrust = 76,160 lb. or 34 tons, which sounds a great deal, but this pressure is spread over the whole area of the cage's base, which is πr^2 or $3{\cdot}14159 \times 12{\cdot}2^2 = 467$ sq. in. The pressure per square inch is

$$\frac{76{,}160}{467} = 163 \text{ lb.}$$

a respectable figure, but not enormous. Nevertheless the cage and the rest of the framework must be made of stout materials; it can be seen that a burst at such pressures could cause pieces to fly off dangerously. The pressing of grapes by the amateur can be done slowly and thus a better quality must be obtained. To see the juice gushing from the press by rapidly tightening it is a temptation to be resisted. 'Slow and steady' is the rule.

A wine-press can, of course, be purchased and cider-presses can be adapted to process grapes. There is even a small model of the famous French Vaslin presses available. This is the 3-hectolitre (66-gallon) horizontal press, holding about three-quarters the quantity of the cage described above. The principle of the Vaslin presses, much used by professional wine makers, is that a horizontal fibreglass cage, filled with grapes, rotates while two internal plates advance towards each other (one movable plate in the small presses), thus crushing the fruit. Stainless steel chains and rings between the two plates collapse as the plates approach each other and tighten as they retreat, thus breaking up the pomace ready for another pressing. Two electric motors supply the power. These presses give excellent results, and can be obtained from Mr. Anton Massel, Ockley, Surrey. The small model is a very neat little machine, 5 ft 3 in. long by 3 ft 1 in. high.

The amateur vigneron without a centrifuge for cleansing the must will find the *guillage* system useful (see page 134). This is fermenting in a full vessel, such as a barrel on its side. Under these conditions odd pips, skins, spent yeast and dirt are thrown out as the fermentation proceeds. As the gas and debris escape, the level of the must in the vessel falls and the container or vessel must be topped up every day from a reserve supply of must. Alternatively clean, insoluble, inert objects, such as washed, rounded flints, may be dropped into a barrel to raise the level of the liquid in it. Perfectly splendid objects, which can be used to meet an emergency, are the children's glass marbles, but here caution is needed. The possible resulting family crisis might become much more serious than the drop in the must level in the barrels!

11

Finale

The commercial production of wine from English or Welsh-grown grapes is now well under way. It is hard work and calls for both devotion to the task and patience in acquiring knowledge. A feature of these vintages is that they are 'single vineyard' wines, each one having its own special characteristics, and as a consequence they are very superior to the standardized bulk wines produced by the wine-growing countries in 10,000-gallon vats and shipped in tankers, like oil. English wines are far from being 'ordinaire' and must be a little more expensive as a result. In skilled hands the wines produced are excellent: numerous blind tastings have shown the white wines to be fully equal to similar German and French vintages. For instance, at the competition held in Christie's Great Rooms, London, in September 1975, to present the Gore-Browne trophy for the best English wine (won by Mr. N. de M. Godden, Pilton Manor, Somerset), six English and six Continental wines were tasted blind by the some 200 people present. The score cards showed that that particular audience at any rate thoroughly approved of the Englanders. In May 1976, as a prelude to a summer exhibition, British Rail organized a tasting of British wines at the Charing Cross Hotel. Twenty-five wines were shown to a professional audience of tasters and personalities in the wine trade and many were the surprises registered by the experts on the range and quality of those bottles.

English wines are now accepted by the public and are no longer considered to be a joke or a curiosity. There are at least twenty-five on the market and leading restaurants and wine merchants now have them on their lists. They are found on board the *Queen Elizabeth II*.

Of course, in most seasons English must has to be sugared to bring it up to a strength giving 10 per cent alcohol, but this is no drawback provided the addition is reasonable. Sugaring is done in France and Germany and is legal there within certain limits, as is also acidity reduction.

At present the financial rewards to the English vigneron are not great, but money need not be lost in the establishment of his vineyard. The unfortunate thing for this new, promising industry is that the English grower using English grapes has been singled out by the government (that is, the Treasury) for greater punishment than the producer of wine from imported grape concentrates coming mostly from Spain and Yugoslavia, countries not even in the Common Market. It is greatly to be hoped that this anomaly can be abolished and that the home-grown material will no longer be taxed at a higher rate than the imported product. The figures on page 143 will bear repetition here. A table wine of 10 per cent alcohol content, made from imported pulp pays an excise tax of £2·11 per gallon: a similar (actually a better) wine made from fresh English grapes pays £3·25 per gallon—54 per cent more. Why? Furthermore, owing to the ingenious arrangement of paying a tax on a tax, Value Added Tax of 8 per cent must be paid on both products, so that the share of tax on tax for the imported material is 16·9p and for home-grown wines is 27p, further punishing the home grower. 'Viva el rey Juan Carlos', and do not forget President Tito as well. The English vigneron might well wonder why we favour them: we can only assume that teacher knows best. The quantities of grape concentrate imported are considerable. As Alexander Pope put it:

> '*In spite of pride in erring reason's spite*
> *One truth is clear* WHATEVER IS IS RIGHT.'

A method of overcoming this heavy excise duty has been put to me by someone who had better remain anonymous for the moment, though it appears to be quite legal. Perhaps it would have to be tested in the courts.

This 'Rent a Vine' scheme is simplicity itself. You rent a vine, or several vines, from a vigneron for an agreed price, say a dozen vines for £13 per annum. The grower looks after the vines for you (you can come and inspect them during the season, of course, and tie a label on them too if you wish), he picks the grapes, makes the wine and bottles it. As the vines are yours all the time you pay rent for them and as the wine from them will be drunk by your yourself, or your friends, and none of it sold it should carry no duty. The vigneron would guarantee you at least one bottle of finished wine for each vine rented (the quantity guaranteed would vary according to the number of vines growing per acre). You would thus get a case of excellent wine for £13, saving yourself £6·60 in duty and purchase tax. As the vigneron in looking after your vine would be supplying you with a service you might have to pay 8 per cent purchase tax on the service element in the transaction, say 8 per cent on half the rental, that is 52p per case, which would be much less than 8 per cent on the full price including duty. The scheme seems to have infinite possibilities for regular customers.

To grow your own wine, to sit in your own vine arbour, then to talk with your friends there or to drink your own vintage sitting round a winter fire is most satisfying.

> '*In her [Queen Elizabeth I's] days everyman shall eat in safety,*
> *Under his own vine, what he plants; and sing*
> *The merry songs of peace to all his neighbours.*'

That was an ideal in Shakespeare's day, that could more easily be realized today.

APPENDIX I

Weather records June–September
Averages of thirty years

Place and Month		Temperature average Max °C	Min °C	Precipi- tation mm	Sunshine Hours	% of poss.	No. days with no sun
Sandown, I.O.W.	June	24·7	7·3	40	254	51	1
(50° 39′N)	July	25·0	10·1	56	236	47	1
	Aug.	25·1	9·6	60	223	49	1
	Sept.	22·7	6·6	66	168	44	2
Total		97·5	33·6	222	881	—	5
Clacton-on-Sea	June	23·9	6·1	41	231	47	1
(51° 47′N)	July	25·6	8·9	47	212	43	1
	Aug.	24·4	8·9	48	199	44	1
	Sept.	22·8	6·7	50	161	42	2
Total		96·7	30·6	186	803	—	5
Rheims, France.	June	30·2	4·7	51	219	45	–
(49° 18′N)	July	32·5	7·3	57	221	45	–
	Aug.	31·2	6·7	63	201	45	–
	Sept.	27·9	2·8	58	165	43	–
Total		121·8	21·5	229	806	—	–
Frankfurt-am-Main, Germany	June	31·4	7·4	73	227	47	1
(50° 7′N)	July	32·7	10·1	70	228	47	1
	Aug.	32·2	9·1	76	204	46	1
	Sept.	28·3	4·7	57	146	38	2
Total		124·6	31·3	276	805	—	5

Source: *Tables of temperature, relative humidity, precipitation and sunshine for the world.* Part III. Europe and the Azores. (1973), London, Meteorological Office 0.856C. HMSO.

APPENDIX II

Sugar and alcohol

The weight of strained grape juice relative to an equal volume of water—that is, the specific gravity—will inform one of the amount of sugar in that juice and the potential alcohol content of the wine made from it. A number of sugar/alcohol tables are available, showing minor discrepancies in the figures. For instance, at 80° Oe the following alcohol contents are given by different authors: Hallgarten 10·6 per cent, Chancrin 10·8 per cent, Pearkes 10·5 per cent, Ordish 10·8 per cent.

The discrepancies arise from varying quantities being allowed for unfermentable sugar and debris in the must.

The table below is based on Chancrin.[20]

The gravities are in ° Oeschler; for an explanation of this system see page 121. The sugar content is given in lb. per 100 gallons, which is the same as grammes per litre. Two additional columns are added showing the sugar needed to be added to the must to raise the alcohol content to 10 per cent and 12·5 per cent (by volume) respectively.

Gravity of must °Oe.	*Sugar lb. per 100 gallons and grammes per litre*	*Potential alcohol % by vol.*	*Add sugar per 100 gall. To increase alcohol to 10% lb.*	*to 12½% lb.*
50	103	6·0	68	112
52	108	6·3	63	107
54	114	6·7	56	101
56	119	7·0	51	96
58	124	7·3	46	91
60	130	7·6	41	85
62	135	7·9	36	80
64	140	8·2	31	75
66	146	8·6	24	69
68	151	8·9	19	64
70	156	9·2	13	59
72	162	9·5	8	53
74	167	9·8	3	48
75	170	10·0	—	45
76	172	10·1	—	43

Appendix II

Gravity of must °Oe.	Sugar lb. per 100 gallons and grammes per litre	Potential alcohol % by vol.	Add sugar per 100 gall. To increase alcohol to 10% lb.	to 12½% lb.
78	178	10·5	—	37
80	183	10·8	—	32
82	188	11·0	—	27
84	194	11·4	—	21
86	199	11·7	—	16
88	204	12·0	—	11
90	210	12·3	—	5
91	212	12·5	—	—
100	236	13·9	—	—

EXAMPLE

If the gravity of the must is 72° Oe it has 162 lb. of sugar per 100 gallons and will make a wine having about 9·5 per cent of alcohol by volume. If the vigneron wishes to make a 10 per cent wine he should add 8 lb. of sugar per 100 gallons of must. If he is aiming at a 12½ per cent wine, he should add 53 lb. of sugar per 100 gallons.

If you prefer to work in metric measurements remember that lb. per 100 gallons and grammes per litre are the same.

APPENDIX III

The English Vineyards Association

List of members, Nov. 1976

CORPORATE MEMBER

Merrydown Wine Co. Ltd., Horam Manor, Horam, Heathfield, Sussex

FULL MEMBERS

S. Alper, Chilford Hall, Linton, Cambridge
B. T. Ambrose, Nether Hall, Cavendish, Sudbury, Suffolk
C. M. D. Ann, Valley Wine Cellars, Drusilla's Corner, Alfriston, Sussex
Lt.-Com. P. Baillie-Grohman, Lodge Farm, Hascombe, Godalming, Surrey
K. C. Barlow, Locks Cottage, Adgestone, Sandown, I.O.W.
R. A. Barnes, Biddenden Vineyards Ltd., Little Whatmans, Biddenden, Ashford, Kent
Rev. D. L. Barrett, 33 Lea Bank Avenue, Kidderminster, Worcester
Mrs. I. M. Barrett, The Vineyards, Crick's Green, Felsted, Essex
J. G. Barrett, The Vineyards, Crick's Green, Felsted, Essex
T. Bates, Cherry Hill, Nettlestead Green, Wateringbury, Kent
K. G. Bell, Thornbury Castle, Thornbury, Bristol
I. H. Berwick, Broadwater, Framlingham, Suffolk
C. R. J. Cadogan-Rawlinson, Brampton Hall, Brampton, Beccles, Suffolk
W. L. Cardy, Lower Bowden, Pangbourne, Berkshire RG8 8JL
D. Carr-Taylor, Yew Tree Farm, Westfield, Hastings, Sussex
J. A. Cave, Paynetts Oast Farm, Goudhurst, Kent
C. Clark, Cherry Bank Estate, Otley, Ipswich, Suffolk
P. W. Cook, Mill Lane Farm, Pulham Market, Diss, Norfolk
J. M. B. Corfe, Ashwell, Ivy Hatch, Sevenoaks, Kent
N. C. Cowderoy, Rock Lodge, Scaynes Hill, Haywards Heath, Sussex RH17 7NG
J. C. Crossland-Hinchcliffe, Castlehouse, Plumpton Green, Lewes, Sussex

J. H. Daltry, Chevel House, The Belt, South Kilworth, Lutterworth, Leicestershire LE17 6DX
R. S. Don, Park House, Elmham, Dereham, Norfolk
P. Dow, The Priory, Brandeston, Woodbridge, Suffolk
J. T. Edgerley, Manor Farm House, Kelsale, Saxmundham, Suffolk
H. B. Evans, Vine Cottage, Park Place, Arundel, Sussex
K. Fitzgerald, Suffolk Vineyards Ltd., Cratfield, Halesworth, Suffolk
Sir Charles Forte, 166 High Holborn, London WC1V 6PF
N. Foster, Lescham Hall, Kings Lynn, Norfolk
C. E. George, 13 Mountain Ash Close, Colchester, Essex
D. Gibbs, Braishfield Manor, Romsey, Hampshire
N. de M. Godden, The Manor House, Pilton, Shepton Mallet, Somerset
Miss P. Gommes, Rake Manor, Milford, Surrey
Mrs. A. M. Gore-Browne, The Vineyards, Beaulieu, Hampshire
S. W. Greenwood, New Hall, Purleigh, Essex
P. A. I. Hall, Breaky Bottom Farm, Northease, Lewes, Sussex
Dr. S. Henry, Palmers Close, Hilperton, Wiltshire
Miss C. Hitchcock, Snorscomb Mill, Everdon, Northamptonshire
G. Jackson, Frensham Manor Vineyards, c/o Pink and Arnold, Wickham, Fareham, Hampshire
R. W. M. Keeling, Monier-Williams & Keeling, 1 Vintner's Place, Upper Thames Street, London EC4V 3BQ
J. D. Knight, Hillside Cottage, Poulner, Ringwood, Hampshire
P. G. Latchford, 38 Crouchfield, Boxmoor, Hemel Hempstead, Hertfordshire
K. McAlpine, The Priory, Lamberhurst, Tunbridge Wells, Kent
I. W. M. MacFarlane, Elder Hall, Peasenhall, Saxmundham, Suffolk
M. I. G. MacKinnon, Corner House, Allins Lane, East Hendred, Berkshire
G. Marsden, Scriven Park, Knaresborough, Yorkshire HG5 9DF
A. Massel, 'Woodlands', Hazel Grove, Hindhead, Surrey
J. Milkovitch, Eastfield House, East Harling, Norfolk NOR 12X
A. J. Miller, Allerton Hill, Windlesham, Surrey
Belinda, Lady Montagu, Kings Row, Blackfield, Southampton, Hampshire SO4 1XR
Lord Montagu of Beaulieu, Palace House, Beaulieu, Hampshire
I. R. Paget, The Old Station House, Singleton, Chichester, Sussex
T. B. L. Parker, 19 The Cliff, Roedean, Brighton, Sussex
Miss G. Pearkes, Yearlstone House, Bickleigh, Devon
W. B. N. Poulter, 27 Broadlands Avenue, Shepperton, Middlesex
G. P. Reece, The Vineyard, Drove Road, Gamlingay, Sandy, Bedfordshire
T. Ridley, Phalaeonopsis Limited, Tilgates, Bletchingley, Surrey

Major Alan Rook, Stragglethorpe Hall, Lincoln
Maj.-Gen. Sir Guy Salisbury-Jones, Mill Down, Hambledon, Hampshire
D. Simmons, Little Pook Hill, Burwash Weald, Sussex
Capt. R. G. Sheepshanks, Rookery House, Eyke, Woodbridge, Suffolk
A. Skuriat, 55 Musters Road, Ruddington, Nottingham
N. J. Sneesby, 17 Luard Road, Cambridge CB2 2PJ
C. D. Suter, Drews Farm, Rowlands Castle, Hampshire
B. H. Theobald, Westbury Farm, Reading, Berkshire
Dr. G. I. Thomas, 16 Glenwood Avenue, Bognor Regis, Sussex
R. D. Thorley, Reyson Oasts, Broad Oak, Rye, Sussex
C. C. G. Trump, Burrow Farm, Broadclyst, Exeter, Devon EX5 3JA
Miss I. C. Vaugham-Morgan, Warren Lodge, Finchampstead, Berkshire
Sir Derek Vestey, 5 Carlton Gardens, London S.W.1.
J. L. Ward, Merrydown Wine Co. Ltd., Horam Manor, Horam, Heathfield, Sussex
M. Waterfield, The Wife of Bath, 4 Upper Bridge Street, Wye, Ashford, Kent
P. Tyson-Woodcock, Casa Nova, 50020 Panzano in Chianti, Prov. Firenze, Italy
W. J. Woodward, Sandfield House, Potterne, Devizes, Wiltshire
Major A. H. Yates, Roughters, Icklesham, Sussex

ASSOCIATE MEMBERS

J. Abbs, Old Thatch, Bistock, Doddington, Sittingbourne, Kent
F. D. Adams, Crantock, 34 Oaktree Lane, Cookhill, Alcester, Warwickshire
R. C. Aikman, Holly Farm, The Heywood, Diss, Norfolk
P. Anderson, Upper Nash Farm, Nutbourne Lane, Pulborough, W. Sussex
A. S. Antoniazzi, 222 Merthyr Road, Pontypridd, Glamorgan
C. Armitage, 6 Marston Avenue, Chessington, Surrey
M. Ashbolt, Wood Farm, Wicklewood, Wymondham, Norfolk
W. T. Ashe, Church Farm, Staple, Canterbury, Kent
The Hon. John Astor, MP, Kirby Farm Office, Inkpen, Berkshire
R. L. Atkins, 35 Preston Park Avenue, Brighton, Sussex BN1 6HE
L. B. Aunger, Melton Cottage, Charlton, Radstock, Bath
C. E. Ayers, Wash Farm, Furneaux Pelham, Buntingford, Hertfordshire
M. Ayers, Suffolk Vineyards Ltd., Cratfield, Halesworth, Suffolk
R. M. Bache, The Crundels, Astley, Stourport on Severn, Worcestershire
A. A. Baker, Alpina, Theescombe Lane, Amberley, Stroud, Gloucestershire GL5 5AS

H. A. Baker, c/o R. H. S. Garden, Wisley, Ripley, Woking, Surrey
M. W. McL. Baker, The Pant Farm, Cross Ash, Abergavenny, Monmouthshire
R. W. Baker, Tumblers Cottage, Thursley Road, Elstead, Surrey GU8 6DH
J. M. Baldock, Hollycombe House, Liphook, Hampshire
G. C. Barcley, Whitstone House, Bovey Tracey, South Devon
Dr. Barnes, Chesterford Park Research Station, Fisons Agrochemical Division, Saffron Walden, Essex
S. Barratt, 245 Haverstock Hill, London N.W.3.
G. M. Barris, 12 Sherbrook Hill, Budleigh Salterton, Devon
A. Barter, Revells Farm, Linton, Ross-on-Wye, Herefordshire HR9 7SD
J. Baruzzi, 1 Oakley Street, Semaphore Park, 5019, South Australia
R. P. Basham, Snipe Cottage, Snipe Farm Road, Clopton, Woodbridge, Suffolk
C. A. Bass, Cragg House Cottage, Snelson Lane, Marthall, Knutsford, Cheshire
D. S. Bates, 41 Grenfell Road, Stoneygate, Leicester
D. R. Baylis, North Heath Farm, Chieveley, Newbury, Berkshire
Dr. F. W. Beech, University of Bristol, Dept of Agriculture & Horticulture, Long Ashton Research Station, Bristol
A. G. Bell, Elmham House, North Elmham, Dereham, Norfolk
V. G. Bell, Keepers Cottage, The Heath, North Elmham, Dereham, Norfolk
D. R. Belman, Sabell & Co. (Birmingham) Ltd., Saxon Way, Birmingham B37 5AX
P. J. Beswick, 2 Calf Lane, Stanton St. Bernard, Marlborough, Wiltshire
J. L. M. Bevan, Croffta, Groes-Faen, Pontyclun, Glamorgan CF7 8NE
A. H. Bingham, Soranks Manor, Fairseat, Sevenoaks, Kent
H. I. Bird, Beech Knoll, Poulner Hill, Ringwood, Hampshire BH24 3HR
J. A. Blackman, 83 Hungerdown Lane, Lawford, Manningtree, Essex CO11 2LY
Staff-Sgt. A. C. Blay, c/o First Cottage, Padbury, Buckingham, Buckinghamshire
R. H. Blayney, Elms Farm, St. Mary, Jersey, Channel Islands
J. F. Bolam, 32 Betham Road, Breenford, Middlesex
K. E. R. Bolton, The Green Farm, Page's Green, Wetheringsett, Stowmarket, Suffolk
Count Bonacassi, Lungarno A, Vespucci 28, Florence, Italy
Maj.-Gen. H. M. C. Bond, Moigne Combe, Warmwell, Dorchester, Dorset

L. J. Bowden, 59 Watson Road, Rotherham, South Yorkshire S61 1JS
A. K. Bowley, Church Farm, Ashton Keynes, Swindon, Wiltshire
D. O. G. Breton, 12 Thurloe Square, London SW7 2TA
A. Bristow, The Grange, Thwaite, Eye, Suffolk IP23 7EE
E. Britnell, Town Farm, Beacon Hill, Penn, High Wycombe, Buckinghamshire
Col. J. J. Broad, 3 Bridstow Place, London W.1.
J. M. Broadbent, Christie's, 8 King Street, St. James's, London SW1Y 6QT
F. J. Brooks, 12 Crescent Road, Wokingham, Berkshire RG11 2DB
N. Brown, 4 Water Lane, Melbourn, Royston, Hertfordshire SG8 6AY
P. E. Brunning, 7 Meadowfield Road, Sawston, Cambridge CB2 4HS
H. R. F. Burr, Priors Farm, Stoke, Andover, Hampshire SP11 OEU
A. J. Butler, Rose Cottage, Park Corner, Nettlebed, Oxford
Mrs. E. P. Butler, The Honeypot, 305 Old Birmingham Road, Lickey, Bromesgrove, Worcestershire
R. P. Butler, Clyst House, Clyst Honiton, Exeter, Devon
K. Bywater, Flat 19, Bermuda Court, 11 Derby Road, Bournemouth BH1 3PY
Dr. G. L. Caldow, Barn Cottage, Stert, Devizes, Wiltshire SN10 3JD
R. J. Campbell, Beerland Farm, Ryall, Bridport, Dorset DT6 6EJ
R. M. O. Capper, The Stocks Farm, Suckley, Worcestershire
W. Carcary, Mill Down Cottage, Hambledon, Hampshire PO7 6RY
J. D. Carter, 21 Winfield Avenue, Brighton BN1 8QH
D. J. Cater, 21 Newstead Way, Wimbledon, London S.W.19
M. J. P. Chilcott, 20 Bourne Avenue, Salisbury, Wiltshire
R. W. Chisholm, Ringden Wood, The Mount, Flimwell, Wadhurst, Sussex TN5 7ON
F. B. Christian, Reve's Place, Wing, Oakham LE15 8SD
Mrs. H. I. Clemons, FCCA, 116 High Street, Harlington, Middlesex UB3 5AD
G. D. Clark, Guild of Sommeliers, Southern Area Branch c/o 49 Hertford Road, Brighton, Sussex
C. E. B. Clive-Ponsonby-Fane, Brympton D'Evercy, Yeovil, Somerset
Mrs. S. C. Coates, Sudburys Farm, Little Burstead, Billericay, Essex
J. H. M. Cockayne, Creems, Nayland, Colchester, Essex
E. V. M. Colegrave, Dowgate, Sandhurst, Hawkhurst, Kent
H. E. Collier, 23 Crescent Road, Beckenham, Kent BR3 2NF
G. B. Cooper, Stile Educational Co., Beaumont Close, Banbury, Oxfordshire OX16 7RG
R. D. Cooper, Oak Cottage, The Hurst, West Peckham, Maidstone, Kent
S. R. Copley, The Old Rectory, Little Henney, Sudbury, Suffolk
K. Cornwell, The Little Priory, Sandy Lane, Nutfield, Surrey

C. T. Courtney-Lewis, Long Hazard, Butlers Cross, Aylesbury, Buckinghamshire HP17 OXH
Dr. H. H. Crabb, Southcott House, Pewsey, Wiltshire
J. G. Cridlan, Croft Hollow, Wood Drive, Sevenoaks, Kent TN13 2NL
M. W. Crisp, Hill Farm, Boyton End, Baythorne End, Halsted, Essex
Dr. R. W. Crocket, Garlands, West Bergholt, Colchester, Essex
G. B. Crosthwaite, Grove Hill, Suckley, Worcestershire
P. J. Crowe, Polmassic Vineyard Ltd., Kerensa, 3 Chute Road, Gorran Haven, St. Austell, Cornwall
R. Daglish, Gatehouse, Crown Lane, Ardleigh, Essex
Mrs. J. Davies, Merriments Farm, Hurst Green, Sussex TN19 7RQ
K. E. Davies, 4 Blenheim Street, New Bond Street, London W.1.
J. Dawson, 53 Perryfield Way, Ham, Richmond, Surrey TW10 7SL
P. H. Day, Chickering Hall, Hoxne, Diss, Norfolk
G. S. Dean, Caldey View Farm, Barriets, Llanteg, Narberth, Dyfed
M. Dean, 12 Richmond Crescent, London N.1.
W. R. Dixon, Yew Tree Cottage, Glascoed, Pontypool, Monmouthshire
H. A. C. Dod, Hazelbourne, Ermyn Way, Leatherhead, Surrey
Mrs. D. Donald, The Garden House, West Tytherley, Salisbury, Wiltshire
P. Dow, The Priory, Brandeston, Woodbridge, Suffolk
Cdr. E. G. Downer, 19a High Street, Portsmouth, Hampshire PO1 2LP
A. W. Downs, 12 Wesley Avenue, Rhoose, Barry, South Glamorgan
Miss M. L. Duck, C.E.S. (Overseas) Ltd., Bridge House, 181 Queen Victoria Street, London E.C. 4
G. Duncan, 12 Palatine Road, Withington, Manchester M20 9JH
J. C. Dunkley, Podere Riecine, Gaiole-in-Chianti (Siena), Italy
R. J. Durling, 3 Bearwood Road, Barkham, Wokingham, Berkshire
J. D. Edeleanu, Haydown, Great Buckland, Ludderdown, Gravesend, Kent
B. R. Edwards, Two Ways, Salters Hill, Tewkesbury, Gloucestershire
G. F. Eisele, 335 Farnham Road, Slough, Buckinghamshire
E. McC. Elliott, Great Thorndean Farm House, Warninglid, Haywards Heath, Sussex
Dr. John Elliott, Spenithorne, Hambleden, Henley-on-Thames, Oxfordshire
F. C. Ernst, High Birch Poultry Farm Ltd., Weeley Heath, Clacton, Essex
Mrs. S. M. Evershed, The Garth House, Tillington, Petworth, Sussex
S. V. H. Fage, 42 Woodfield Avenue, Farlington, Portsmouth, Hampshire

M. Farrell, Preston Hill Farm, Wootton Wawen, Warwickshire
R. F. Farrer, Wye College, Wye, Ashford, Kent (Hon. Member)
C. D. Farrier, 110 Ragged Hall Lane, St. Albans, Hertfordshire
G. N. Farrow, 2 Oakwood Cottages, Shipley, Horsham, Sussex
H. S. S. Few, Oakington House, Oakington, Cambridge
E. J. Fitch, 14 Riversmeet, Hertfordshire
J. Fitzgerald, 1011 Oxford Road, Tilehurst, Reading, Berkshire RG3 6TL
J. Fletcher, Cheltra, Park Drive, Claverdon, Warwickshire
Dr. A. E. Forbes, Elmfield House, Fordton, Crediton, Devon EX17 3DH
A. Freedman, The Old Rectory, Bucknell, Bicester, Oxfordshire OX6 9LT
D. J. Freeman, Leicester Grange Farm, Hinckley, Leicester
T. M. J. French, Higher Farm, Stoke-in-Teinhead, Newton Abbot, Devon
A. Friend, Lower Bromley Coombe Farm, Stoke Abbott, Beaminster, Dorset DT8 3JZ
Mrs. Fuller, Wakelins, Genesis Green, Wickhambrook, Newmarket, Suffolk
H. E. Fuller-Lewis, Marketing Manager, Agrochemical Division, Ciba-Geigy (UK) Ltd., Whittlesford, Cambridgeshire CB2 4QT
J. Fulton, 158 Abbots Road, Abbots Langley, Hertfordshire
M. Furness, 16 Grafton Square, Clapham, London S.W.4.
G. Garbutt, c/o 148 Sloane Street, London S.W.1.
E. I. Garratt, Dunsley House, Kinver, Stourbridge, Worcestershire DY7 6NB
E. Gay, The Down Wood, Blandford Forum, Dorset DT11 9HN
F. F. George, Wixford Lodge, Bidford-on-Avon, Alcester, Warwickshire
S. E. Gerrard, Yerba Buena, New Domewood, Copthorne, Sussex
R. Gibb, 1 Park Avenue, Madeley, Telford, Shropshire TF7 5AB
R. Gibbons, The Mitre Restaurant, 56 High Street, Old Town, Hastings, Sussex
R. H. Gibbons, Cranmore Vineyard, Solent Road, Cranmore, Yarmouth, Isle of Wight
D. M. Gilbertson, Wethers, Fyfield, Essex
Major C. L. B. Gillespie, North Town House, North Wootton, Shepton Mallet, Somerset
S. C. J. Gilling, Tai-Lu, Coronation Road, Kington, Herefordshire HR5 3BU
A. H. Goddard, Chancers, Ampney St. Mary, Cirencester, Gloucestershire GL7 5SN
F. M. Godwin, 85 Morland Road, Croydon, Surrey

Miss J. Goodman, 7 The Fields, Tacolneston, Norfolk NR16 1DG
M. D. C. Goodridge, Stickeldown, 27 Poulner Close, Felpham, Bognor Regis, Sussex
A. H. Goodwin, 16 Castle Street, Ruthin, Clwyd
C. P. Gordon, Hoe House, Peaslake, Guildford, Surrey GU5 9SR
I. A. Grant, Knowle Hill Farm, Ulcombe, Maidstone, Kent
J. Neville Grant, Dept. of Liberal Studies, College of Food Technology and Commerce, Colchester Avenue, Cardiff CF3 7XR
R. A. Graves, 12 The Drift, Oakington, Cambridge CB4 5AD
K. R. Gray, Lymington, 30 Upland Road, Selly Park, Birmingham, Warwickshire B29 7JS
B. R. Grose, Flat 5, 36 Southend Road, Beckenham, Kent
Mrs. L. Grover, Garden Cottage, Rookery Road, Downe, Kent
Milton Grundy, Gray's Inn Chambers, Gray's Inn, London W.C.1.
The Hon. Charles Guest, Grove Farm, Tolland, Taunton, Somerset
P. Guest, Ivy Farm, Grimston, King's Lynn, Norfolk
P. N. Guilford, Kalkara, Station Road, Deeping St. James, Peterborough, Northamptonshire
W. S. C. Gurney, Yew Tree Farm, Silver Green, Hempnall, Norwich, Norfolk NOR 64W
Dr. P. A. Hallgarten, c/o S. F. & O. Hallgarten, Highgate Road, Carkers Lane, London NW5
F. F. Hamments, 40 Chestnut Drive, Harrow Weald, Middlesex
E. H. Hanson, 9 Dale Hall Lane, Ipswich, Suffolk IP1 3RX
D. St. C. Harcourt, Bassetts, Mark Cross, Crowborough, Sussex
C. S. Harker, Holly Dell, Satwell Close, Rotherfield Greys, Henley-on-Thames, Oxfordshire
R. A. Harmer, Whitehall, Compton Abbas, Shaftesbury, Dorset SP7 OLZ
J. S. Harper, 5682—184th St., Surrey B.C., Canada
Sir William Hart, Turweston Lodge, Brackley, Northamptonshire.
P. J. Harvey, 239A Springfield Road, Chelmsford, Essex CM2 6JT
M. Hasslacher, Deinhard & Co. Ltd., Deinhard House, 29 Addington Street, London SE1
D. G. Headley, Fulbrook Cottage, Blockley, Moreton-in-Marsh, Gloucestershire
P. J. Heagerty, Marchants Farm, Streat Lane, Streat, Hassocks, Sussex BN6 8RY
A. G. D. Heath, 'Glenlyn', 54 Lakewood Road, Chandlers Ford, Hampshire SO5 1EX
R. O. Hender, 10 Red House Lane, Westbury-on-Trym, Bristol BS9 3RY
B. D. Henderson, Dalingridge Farm, Sharpthorne, East Grinstead, Sussex

H. H. Hills, 13 Church Road, Burmatsh, Romney Marsh, Kent
G. C. Hilscher, 6 Kewferry Dr., Northwood, Middlesex HA6 2RM
B. J. Hoar, 4 Tylers, Sewards End, Saffron Walden, Essex
G. B. Hole, Frogmore Farm, Bradfield, Berkshire
A. S. Holmes, Vine Lodge, Wraxhall, Shepton Mallet, Somerset BA4 6RQ
B. B. Honess, Gt. Cheveney House, Marden, Tonbridge, Kent
M. R. Hope, Barningham Hall, Barningham, Bury St. Edmunds, Suffolk
D. V. Hopkins, 29 Lon Cae Porth, Rhiwbina, Cardiff, Glamorgan
J. R. Houghton, Hillfarrance, Cropwell Road, Langar, Nottinghamshire
C. G. C. Houry, Lowicks House, Tilford, Surrey GU10 2EX
A. H. Hubbard, Laundry Cottage, Serlby, Doncaster, South Yorkshire DN10 6BA
P. Hunt, Higher Yellands, Whimple, Exeter, Devon
A. Hunter, 16 Longleat Square, Farnborough, Hampshire
M. Hurle, 353 Quemerford, Calne, Wiltshire
K. J. Ingram, 1 Orchard Close, Hagley, Stourbridge, West Midlands
K. R. H. James, Broadfield Court, Bodenham, Herefordshire
R. Jeffries, Willows, Duddenhoe End, Saffron Walden, Essex
R. G. Jenkins, Virgins, West Hatch, Taunton, Somerset
E. M. H. Johnson, Halford Cottage, Halford, Shipton-on-Stour, Warwickshire
H. G. Jones, 2 Little Orchard, Dinas Powis, Glamorgan, South Wales CF6 4NH
R. P. Jones, Kennet View, Avon Way, Bath Road, Padworth, Reading, Berkshire RG7 5HR
T. M. Jones, The Hollies, 11 Park Road, Nantwich, Cheshire
E. P. Kelly, Agricultural Institute, Ballygagin, Dungarvan, Waterford, Ireland
J. D. Kelly, Keepers, Norwood Lane, Graffham, Petworth, West Sussex GU28 OQQ
P. W. Kemmis, Hangar Farm, Cornwood, Ivybridge, Devon
G. H. Kinch, Canda Gates, 1 Hulbert Road, Bedhampton, Havant, Hampshire
P. A. King, The Homestead, Addis Lane, Cutnall Green, Droitwich, Worcestershire
Miss S. King, 28 White Oak Drive, Beckenham, Kent
R. O. Kinnison, Avon Valley Nurseries, South Gorley, Fordingbridge, Hampshire
M. Kirby, 49 Pinewood Green, Iver Heath, Buckinghamshire
J. C. Kitchin, High Street Farm, Boxford, Newbury, Berkshire RG16 8DD

B. W. Knight, The Glen, Langney, Eastbourne, Sussex
M. Knight, 35 Park Avenue, Palmers Green, London N.13
R. M. Knoll, Olivers Orchard Ltd., Olivers Lane, Colchester, Essex
J. F. S. Laidlaw, Stanton, 45 Common Lane, New Haw, Weybridge, Surrey
J. L. Lampitt, Thelsford, Wellesbourne, Warwick
R. E. Lane-Hall, School House, Luddenham, Faversham, Kent ME13 OTE
Col. R. C. Laughton, Kempes Hall, Ashford, Kent
M. J. S. Lawrence, 18 Weetwood Avenue, Leeds, Yorkshire LS16 SNF
W. P. Lawrence, 16 Eriswell Crescent, Walton-on-Thames, Surrey KT12 5DS
G. E. Lawson, 20 St. George's Place, Hurstpierpoint, Sussex BN6 9QT
Dr. J. A. Learner, 56 Friargate, Derby
J. K. Leonard, 60 Nether Street, Bromham, Chippenham, Wiltshire
R. A. Lescott, 10–12 West Hill, Aspley Guise, Milton Keynes
R. Leslie, Glebe House, Langham, Colchester, Essex
P. Levett, Pound Field, Brenchley, Tonbridge, Kent
P. J. Lidgitt, 6 Newlands Park, Dunfermline, Fife KT12 ORG
M. Lilley, 91 Morley Grove, Harlow, Essex
R. G. Ling, Manor Pound, Coleshill, Amersham, Buckinghamshire
T. Lisher, Harvest, High Street, Harlton, Cambridge
M. J. Lockley, 15 Redwood Close, Sutton Coldfield, West Midlands B74 3JQ
J. S. Lodge, 48 Holland Street, London W.8.
J. Lory, John Lory Farmers Ltd., Charlwood Place, Horley, Surrey
L. J. MacDonald, Higher Barn Farm, Stones Green, Great Oakley, Harwich, Essex
N. MacDonald, Batswell, Washfield, Tiverton, Devon EX16 9PE
T. P. McElwee, Thistledown, Fingest, Henley, Oxfordshire
I. W. MacFarlane, Myrtle Nursery, Manallan, Helston, Cornwall
A. McKechnie, Fairfields Fruit Farm Ltd., Rhyle House, Newent, Gloucestershire
G. C. MacLean, Appleby Fruit Farm, Kingston Bagpuize, Abingdon, Berkshire
Miss M. M. P. MacRae, Hatchery House, Barrow, Bury St. Edmunds, Suffolk
W. P. Mansfield, Burscombe Farm, Egerton, Ashford, Kent
H. C. Marks, Vinns Hill, E. Meon, Hampshire
A. Marshall, Old Shields, Ardleigh, Colchester, Essex
Dr. D. M. Martin, CS 1 HU, Stanwell Avenue, Hobart, Tasmania 7,000, Australia
L. A. Mason, Manor Farm, Bury, Huntingdon

Lord May of Weybridge, Gautherns Barn, Sibford Gower, Oxfordshire

J. E. Mayne, Sheepway, Stockwell Lane, Cleeve Hill, Gloucester GL52 3PU

J. M. Mayo, The Bungalow, Waterwells Farm, Quedgerley, Gloucester GL2 6SA

G. C. Meyrick, Laterditch House, Bransgore, Christchurch, Hampshire

Ministry of Agriculture, Efford Experimental Horticulture Station, Lymington, Hampshire SO4 QLZ

Mrs. Monkton, Stretton Hall, Stretton, Stafford

Mrs. S. Moore, Little Court, Highwood Hill, London N.W.7.

Mrs. P. Moran, Bylsborough, Henfield, Sussex

Prof. Dr. h.c. Lenz Moser, 3495 Rohrendorf bei Krems, Austria (Hon. Member)

Herr Laurenz Moser, Lenz Moser GmbH, 3495 Rohrendorf bei Krems, Austria, (Hon. Member)

A. Munro, Wyck Hill House, Stow-on-the-Wold, Cheltenham, Gloucestershire

P. G. Neale, High Cairn, 32 Grange Road, Shrewsbury, Shropshire

H. Norris, 21 Gentleman's Row, Enfield, Middlesex

N. Oakley, Southerndown, Shrivenham Road, Highworth, Wiltshire SN6 7B2

H. A. Ogilvie, Whiteoaks, Battle Road, Hailsham, Sussex BN27 1UE

G. O'Halloran, 10 Pine View Road, Ipswich, Suffolk IP1 4HS

L. A. Oley, Fallowfield, Todds Green, Old Stevenage, Hertfordshire

G. Ordish, 178 London Road, St. Albans, Hertfordshire

G. Padley, c/o Paten & Co., The Maltings, Aldermans Drive, Peterborough

T. B. Paisley, Manor Farm, Holywell, St. Ives, Huntingdon

C. Parkinson, Birchlyn, Redcot Lane, Stenning, Canterbury, Kent

F. B. Patient, 34 Grundisburgh Road, Woodbridge, Suffolk IP12 4HG

C. S. Peace, Harper Lees, Hathersage, Sheffield S30 1BA

M. H. C. Peace, Sheepcote Farmhouse, Wooburn Common, Buckinghamshire

M. R. P. Pelly, c/o Whitbread & Co. Ltd., Beltring, Paddock Wood, Tonbridge, Kent

D. J. Peplow, 5 The Hall, Swindon Village, Cheltenham, Gloucestershire

J. E. Perrin, 21 Alexandra Road, Gloucester GL1 3DR

P. Perry, 22 Applefield, Northgate, Crawley, Sussex RH10 2BJ

C. J. Pettet, 9 Avenue Close, Southgate, London N14 4BJ

Wing-Com. C. J. Phillips, 55 Rounton Road, Church Crookham, Aldershot, Hampshire GU13 OJH

J. L. Phillips, Jeffery Phillips (Wine Merchant) Ltd., 58 South Street, Pennington, Lymington, Hampshire
B. W. Phipps, 9 Upper Hall Park, Berkhamsted, Hertfordshire
A. J. Pinnington, 13 Barrows Croft, Cheddar, Somerset BS27 3BH
I. Piper, 23 Tournay Road, London SW6 7UG
Dr. A. Pollard, The Clearing, Brockley Hall, Backwell, Bristol BS19 3A2
N. H. Porter, Marmings Farm, Small Dole, Henfield, Sussex
L. Potter, Tudor House, Kintbury, Newbury, Berkshire
Mrs. J. M. Pratt, Bookers Vineyard, Foxhole Lane, Bolney, Sussex
Miss C. Preston, 159 Cromwell Road, London S.W.5
Major F. W. Preston, Heatherwode Granary, Buxted, Sussex
Lt.-Col. C. L. Price, Meadlands, Ightham, Sevenoaks, Kent
J. E. Pullinger, 65 Stockton Road, Guildford, Surrey
C. Ramsey, 121 Upper Bristol Road, Weston-Super-Mare, Avon BS22 9AW
R. Reeves, The Priory, Lamberhurst, Tunbridge Wells, Kent
Lord Remnant, Bear Place, Hare Hatch, Reading, Berkshire RG10 9XR
K. J. Reynolds, Pook Reed, Pook Reed Lane, Heathfield, Sussex
A. S. Richardson, 359 Chambersbury Lane, Leverstock Green, Hemel Hempstead, Hertfordshire
P. V. Richardson, Owls Hall, Blackmore End, Braintree, Essex
P. F. J. Rickards, Tuddenham Mill, Bury St. Edmunds, Suffolk
I. P. Le Messurier Ritchie, St. Edmunds Vineyard, Thurston Planche, Bury St. Edmunds, Suffolk IP31 3RQ
R. H. C. Roberts, Bodlondeb, Morfa Nefyn, Pwllheli, North Wales
D. Robinson, Toledo, Sunnymead Lane, Top Sutton, Bishop Sutton, Bristol
G. B. Robinson, 38 Feilden Grove, Headington, Oxford
H. Robson, 30 Hawthorn Drive, Coney Hall, West Wickham, Kent
D. F. J. T. Rodgers, 45 Noddington Lane, Whittington, Lichfield, Staffordshire
C. H. H. Roughton, Cambarn, Dunkerton, Bath BA2 8AS
J. B. Rowlands, Nethercote, Knox Lane, Bardwell, Bury St. Edmunds, Suffolk
H. Royce, Herrings Farm, Cross-in-Hand, Heathfield, Sussex
R. Rushton, Lincoln C.B.C., City Hall, Beaumont Fee, Lincoln LN1 1DD
M. Russell, International Oenological Control, 5 Southview Road, Southwick, Sussex
The Hon. W. S. Russell, Christophers, Newney Green, Writtle, Essex CM1 3SE
E. Saqui, 5 Cologne 51, Lindenallee 59, West Germany

J. D. Scott, 14 Bowers Croft, Cambridge CB1 4RP
Miss Scott-Elliott, Crabadon Manor, Diptford, S. Devon
W. J. Searle, Applegarth, Sion Hill, Bath, Somerset
The Secretary, International Wine & Food Society, Marble Arch House, 44 Edgware Road, London W.2
C. W. Sentance, Furzey Cottage, Wainsford Road, Pennington, Lymington, Hampshire
A. R. Shaw, 41 Trowbridge Road, Bradford-on-Avon, Wiltshire
N. H. Shears, 9 Orchard Road, Westbrook, Margate, Kent
R. J. Sherwood, Ty'r Winllan, Cefnllan, Waun Fawr, Aberystwyth, Dyfed
W. A. Sichel, Winfrith, Misbourne Avenue, Chalfont St. Peter, Buckinghamshire
Dr. C. G. Silcocks, Pitchcombe View, Pitchcombe, Stroud, Gloucestershire GL6 6LW
P. J. Simms, c/o F.C.O. (Freetown), King Charles Street, London SW1A 2AH
A. C. Simpson, Head of Research & Development, I.D. & V's., Gilbey House, Fourth Avenue, Harlow, Essex CM20 1DX
Reresby Sitwell, Renishaw Hall, Derbyshire
S. Skelton, Little Ivy Mill, Loose, Kent
M. C. Skipworth, Lotus Pottery, Stoke Gabriel, Totnes, South Devon
D. H. Sleigh, B & T Group, Sharpestone House, Freshford, Bath BA3 6DA
Lt.-Com. M. C. M. Smith, Downsmead, Pendennis Road, Sevenoaks, Kent
Mrs. P. A. Smith, Flexerne, Fletching Common, Newick, Sussex
R. Smith, Pennavon House, Lower Pennington, Lymington, Hampshire
R. C. Smith, Tanglewood, Little Gaddesden, Berkhamsted, Hertfordshire
W. A. Smith, 9 Roughwood Close, Watford, Hertfordshire WD1 3HN
J. H. Smithes, Coleridge Place, Strete, Dartmouth, Devon
A. Soldani, Well Farm, Church Street, Gt. Burstead, Billericay, Essex
M. W. Stacey, Gorsehill Abbey Farm, Broadway, Worcester
W. J. Stansfield, Field House, Houghton Green, Playden, Rye, Sussex
W. B. C. Stapleton, Newington Grounds, North Newington, Banbury, Oxfordshire
A. Starkey, Coopersfield, West Ashling, Chichester, Sussex
C. L. Steirn, Hook Farm, Broad Common, Worplesdon, Guildford, Surrey GU3 3BL
D. S. Stewart, 143 Pelham Road, Rochester, N.Y. 14610, U.S.A.
J. D. Stewart, Dove-Cote, Killiney Heath, Killiney, Co. Dublin, Eire

D. P. R. Stokes, The Gardens, Sugnall, Stafford ST21 6NF
J. M. Stone, Tredinnick Farm, St. Keverne, Helston, Cornwall
M. A. Stone, Box 3003, Lilongwe, Malawi, C. Africa
R. Streeter, 21 Penlands Vale, Steyning, Sussex
G. G. E. Stroud, 14 Mowbray Road, Sholing, Southampton
C. Stuart, Hailstone House, Cricklade, Wiltshire
R. C. Swan, Timbers, Brassey Road, Limpsfield, Surrey RH8 OET
I. T. Symington, West Winch, King's Lynn, Norfolk
F. Tait, Duphar-Midox Ltd., Smarden, Kent TN27 8QL
J. R. S. Tapp, Wagtail Farm, St. Nicholas-at-Wade, Birchington, Kent
Miss J. C. Denne Tattersall, South Lodge, Ducklys Park, West Hoathly, East Grinstead, Sussex
Miss E. S. Taylor, Casterton House, Casterton Road, Stamford, Lincolnshire PE9 2UA
K. W. B. Thomson, Holly Cottage, Ballycorus Road, Shankhill, Co. Dublin, Eire
D. M. Thorpe, White House Farm, Starston, Harleston, Norfolk
D. K. Tippett, Lamberts Farm, Earls Colne, Colchester, Essex
J. J. Tiranti, Garthfield, Wargrave, Berkshire RG10 SEE
G. C. Todd, 55 Larch Close, Balham, London S.W.12
Dr. N. A. Toes, 42 Maltese Road, Chelmsford, Essex
D. H. G. Tollemache, 76 The Mint, Rye, Sussex
S. G. D. Tozer, York Cottage, Slinfold, Sussex
C. Tunney, Cryals Court, Matfield, Tonbridge, Kent
A. F. Turner, 46 Firle Road, Lancing, Sussex
G. L. Tyler, 2 Walnut Way, Bourne End, Buckinghamshire SL8 5DW
J. S. H. M. Vereker, Orchard Farm, Tutts Clump, Bradfield, Berkshire
P. Vestey, 53 Cheval Place, London S.W.7
J. Vielvoye, B.C. Dept. of Agriculture, 1873 Spall Road, Kelowna, British Columbia V1I 4R2, Canada
D. R. Wagstaff, Heath Engineering Works (Horsmonden) Ltd., Horsmonden, Tonbridge, Kent
R. B. Walker, Weathervane Cott. Troston, Bury St. Edmunds, Suffolk
E. G. Wallis, 45 Swakeleys Drive, Ickenham, Uxbridge, Middlesex UB10 8QD
T. W. Walmsley, Rose Farm, Thorpe-le-Soken, Essex
S. A. Warman, Duffryn, Compton Martin, Bristol BS18 6JA
M. R. Warner, White Acre, White Lane, Guildford, Surrey
A. R. Warren, Deanfield, 9 Latchmoor Way, Gerrards Cross, Buckingham SL9 8LW
J. A. Waters, 14 Glenroyd Gardens, Southbourne, Bournemouth BH6 3JN
D. R. Watkiss, Holme Court, Biggleswade, Bedfordshire
G. Watson, Tara, Woodland Rise, Sevenoaks, Kent

Mrs. I. Way, Bay Tree Cottage, White Leaved Oak, Hollybush, Ledbury, Herefordshire
R. A. Webb, Lincoln Farm Nurseries, High Street, Standlake, Witney, Oxfordshire
A. Webster, 58 Rubislaw Den North, Aberdeen AB2 4AN
W. H. Westphal, The Grove, Penshurst, Kent
J. F. Whitaker, Home Farmhouse, Puttenham, Guildford, Surrey
F. White, 6 London Road, King's Lynn, Norfolk
R. D. White, 14 Frederick Place, Clifton, Bristol 8, Somerset
S. J. Whitley, Haverbrack, Park New Road, Woldingham, Surrey CR3 7DH
B. Wilberforce-Smith, Mill House, Surlingham, Norwich, Norfolk NOR O7W
J. Wilkinson, Shelsley Grange, Shelsley Beauchamp, Worcestershire WR6 6RJ
G. T. Williams, Tasmine, Ratby Lane, Markfield, Leicester
J. G. Williams, Old Barn, Middletown, Studley, Warwickshire
W. A. Williams, Heath Lodge, Arlesford Main Road, Colchester, Essex
G. E. Wiltshire, 5 Caernarvan Street, Launceston, Tasmania, Australia 7250
J. K. Wingfield-Digby, Charlton Barrow, Charlton Marshall, Blandford, Dorset
M. Witt, Coldharbour, West End, Chobham, Woking, Surrey
J. C. Wood, 31 The Lookout, Peacehaven, Sussex
A. Woods, Copper Beech Vineyard, Black Hill, Lindfield, Sussex
D. H. Woods, Flat 13, 35 Kensington Court, London W.8
I. Wrigglesworth, MP, 31 Dunoon Road, London SE23 3TD
J. A. Wright, Monks Staithe, Church End, Princes Risborough, Aylesbury, Buckinghamshire
Mr. Wythe, The Cottage Loaf, 1 Bakers Lane, Codicote, Hertfordshire
C. S. Yee, 16 Oval Road, Camden Town, London NW1 7DJ

APPENDIX IV

Some aspects of modern viticulture in the English climate, 1971

by W. B. N. Poulter.[89]

I am much obliged to Mr. Poulter for allowing me to print the conclusions arrived at from his interesting study. They are:

1. Rather obviously, the nearer a slope approaches south in aspect and the steeper the slope, at least up to 40°, the higher is the insolation during the greater part of the vegetative season.

2. In midsummer when the ecliptic is high the degree of slope has a very small influence on insolation. In spring and autumn however, slope has a significant effect. For example, the difference between the insolation received by a level site and a slope of 30° in a southerly direction is only 8 per cent in midsummer, but in mid October the difference is 70 per cent.

3. Any slope of due westerly or easterly aspect is at all times inferior to a level site in so far as daily insolation summation is concerned.

4. All slope aspects between West-South-West and East-South-East are superior to a level site by way of seasonal insolation summation, although for a short period in midsummer slopes much west of south-west and east of south-east are inferior because of slope shadow in the morning and in the evening respectively.

5. The apparent oddity that a southerly slope of 40° receives less insolation than does a slope of only 30° in midsummer is due to the fact that the sun's incident angle with the slope passes twice through maximum and declines well in the intervening period. In late summer this does not occur, because of the sun's lower altitude.

6. The apparent oddity that a south-west or south-east slope of 40° receives less insolation than a slope of 30° in midsummer is due to slope shadow in morning and evening respectively. In early and late summer this does not occur, because the sun rises and sets further south in azimuth.

7. The direct sunlight duration is greatly affected by slope, aspect and time of the year. For example, (a) in midsummer a westerly or easterly slope of 40° receives 5 hours' less sunlight than does a level site, but in

mid October this difference is reduced to 3·4 hours; (b) in midsummer a southerly slope of 20° receives 2·75 hours' less sunlight than does level ground but in spring and autumn sunlight duration is equal.

8. At all times of the vegetative season direct sunlight is received for the longest duration by a level site. Only in spring and autumn does direct sunlight on near-southerly slopes equal this duration.

9. In midsummer direct sunlight is received for the longest duration by the shallowest slopes, and shallow southerly slopes receive a shorter period of direct sunlight than similar elevations of any other aspect.

10. In spring and autumn southerly slopes of all elevations receive direct sunlight for a longer period than equal elevations of any other aspect.

Bibliography and references

1 Aelfric, *Colloquay*, (10th–11th century) 1628, London.
2 Ashdown, C. A., Roman pavements in Verulamium, *St. Albans Historical Society Transactions*, 1903–14, St. Albans, Gibbs and Bamforth. pp. 166/7.
3 Badell, R. L., Vines européas, *Anales de la Escuela de Peritos Agricolas*, 1952, Barcelona. Vol. XI, pp. 49, 91–232.
4 Barrett, J. G., Spraying—a means to greater profitability, *English Vineyards Association Journal*, 1974. Vol. 8, p. 59.
5 Barry, Sir Edward, *Observations, Historical, Critical on the Wines of the Ancients and the Analogy between Them and Modern Wines*, 1755, London, Cadell. pp. 471–75.
6 Barskii, Y. S., Training bees to pollinate grape vines, *Sad Ogorod*, 1956. Vol. 4, p. 64.
7 Battely, Nicholas, *The History and Antiquities of the Cathedral Church of Rochester*, 1717, London. Appendix p. 27.
8 Bede, the Venerable, *The Ecclesiastical History of the English Nation*, translated by J. Stevens, 1723 (and revised by J. A. Giles, 1847), 1958 edn., London, Dent. pp. 4, 6.
9 *Beowulf*, translated by D. Wright, 1970, London, Panther Books. Line 1162, p. 54.
10 Berkeley, M. J., Oidium Tuckeri, *Gardeners' Chronicle and Agricultural Gazette*, 1845, London.
11 Blickling, *The Blickling Homilies of the Tenth Century*, 1880, London, Early English Text Society. p. 165.
12 Brock, R. Barrington, *Outdoor Grapes in a Cold Climate, Report No. 1*, 1949. *More Outdoor Grapes, Report No. 2*, 1950. *Progress with Vines and Wines, Report No. 3*, 1961. *Starting a Vineyard, Report No. 4*, 1964. Oxted, Surrey, Dunkeld Press.
13 Le Brocq, Philip, *Patent No. 1513*, 1785, Patent Office, London.
14 *A Description with Notes of Certain Methods of Planting, Training and Management of Fruit Trees, Vines, etc.*, 1786, London.
15 Bunyard, G., *The Fruit Garden*, 1904, London, *Country Life*.
16 Burke, E., *Letter to a Member of the National Assembly*, 1906, Oxford, Miscellaneous Works.

17 Camden, William, translated by Edmund Gibson, *Camden's Britannia, newly translated into English with large additions and improvements*, 1695, London. Col. 124, 231, 245, 343, 965.

18 *Canterbury Archives, K. I. 292* (226) Th. 459.

19 Chaboussou, F., La lutte intégrée, autrement dit: la protection de la vigne implique la prise en considération de tous les facteurs susceptibles d'agir sur sa physiologie . . . *Progrès agricole et viticole*, 1974, Paris. Vol. 91, pp. 484–7, 509–17, 580–3, 599–603.

20 Chancrin, E., *Viticulture moderne*, 1950, Paris, Hachette. pp. 31, 223, 245, 297, 319–21.

21 Chaucer, Geoffrey, *The Canterbury Tales*, edited by A. Burrell, 1915, London, Dent. Lines 752, 304, 312.

22 Collier, J., *The Great Historical, Geographical, Geneaological and Poetical Dictionary*, 2nd edn., 1701, London, I. Edwin.

23 *De vino et ejus proprietate*, 1480 (P. Beroalde?), translated (to French) by André Berry, 1939, Paris, Editions Tournelle.

24 *Dictionary of National Biography*, ed. S. Lee, 1909, Pegge, Samuel the elder. Vol. XV, p. 679.

25 Digby, Sir Kenelm, *The Closet of the Eminently Learned Sir Kenelm Digby*, 1669, London.

26 Don, R. S., Marketing and commercial viability of English wines, *English Vineyards Association Journal*, 1974. Vol. 9, pp. 62–7.

27 Duncan, R., Why not grow your own wine? *Evening Standard*, 1 February, 1954, London.

28 Eadwin, *Canterbury Great Psalter* (*c.* 1150), 1905, London.

29 Ellis, Sir Henry, *A General Introduction to the Domesday Book*, 1833, London.

30 European Economic Community, *Regulation No. 925/74 of the Commission*, 17 April, 1974.

31 European Parliament Information Office, *Excise Duty on Wine*, Question No. 411/75, 1975, Brussels.

32 Evelyn, John, *Diary*, 26 September, 1655.

33 Field, J., *English Field-names*, 1972, Newton Abbot, David and Charles. p. 245.

34 Finney, J. R., Farrell, G. M., and Bent, K. J., Bupirimate—a new fungicide, *Proceedings of the 8th British Insecticide and Fungicide Conference 1975*, 1976, Nottingham, Boots. Vol. II, pp. 667–73.

35 Fowler, G. W. A., Poisoned fruit, *Amateur Winemaker*, January, 1975. Vol. XVIII, pp. 1, 8–10.

36 Free, J. B., *Insect Pollination of Crops*, 1970, London, Academic Press. pp. 189–92.

37 Gautier, E.-J. A., *La sophistication des vins*, 3rd edn., 1884, Paris, Baillière. p. 236.

38 Genesis, X: 9.

39 Gnaegi, F., and Dufour, A., Rémanence des fungicides anti-*Botrytis* dans les vins, *Revue suisse de viticulture, arboriculture, horticulture*, 1972. Vol. 4 (3), pp. 101–6.

40 Grainger, J., The Cp/Rs approach to plant pathology, *Span*, 1967, London, Shell. Vol. 10, pp. 1, 44–9.

41 Grison, P., Principes et méthodes de controle intégré, *Academia nazionale dei Lincei*, 1968, Rome. Quad. N. 128, 211–30.

42 Grossmith, G., and Grossmith, W., *The Diary of a Nobody* (1894), 1940 edn., London, Dent. pp. 247, 288.

43 Guyot, J., *Culture de la vigne*, 1860, Paris, Maison Rustique. pp. 30–45.

44 Hallgarten, P. A., Fermentation, *Wine and Spirit Trade Record*, November/December, 1962.

45 Hallgarten, S. F., *German Wines*, 1976, London, Faber. pp. 98, 119, 135.

46 Helgeson, E. A., *La lucha contra las malas hierbas*, 1957, Rome, FAO. pp. 135, 136.

47 *Horticultural Abstracts*, Commonwealth Bureau of Horticulture and Plantation Crops, 1975, Slough. Vol. 45, No. 3084.

48 Hyams, Edward, *The Grape Vine in England*, 1949, London, Bodley Head.

49 Hyams, Edward, Broadcast, BBC (sound), 11 March, 1951.

50 Hyams, Edward, *Dionysius*, 1965, London, Thames and Hudson. p. 187, Plate 70.

51 Hyams, Edward (ed.), *Vineyards in England*, 1953, London, Faber. Plate 2.

52 Jeffs, J., *The Wines of Europe*, 1971, London, Faber. pp. 69–72.

53 Juillard, B., Désherbage et entretien des sols de vignoble, *Progrès agricole et viticole*, 1974. Vol. 91, pp. 827–8. 1975; vol. 92, pp. 11–23.

54 Juvenal, *Satires*. X, 80.

55 Kirby, William, and Spence, William, *An Introduction to Entomology*, 1818, London, Longman. 3rd edn., Vol. 1, p. 205, 1857, 7th edn., p. 115.

56 Lambarde, William, *Perambulation of Kent*, 1576, London. p. 419.

57 Levett, A., *Studies in Manorial History*, 1938, London, Oxford U.P. Appendix.

58 *The Lincolnshire Echo*, Vintage '75 on the way in Lincoln . . ., 12 September, 1975.

59 Malmesburiensis, G., *De Pontificibus* (12th century), 1648, London. Vol. V, p. 131.

60 Marcellinus, Ammianus, Lib. XV, Cambridge, Mass., Harvard U.P.

61 Marletta, G. P., Gabrielli. L. F., and Favretto, L., Lead in grapes

exposed to exhaust gases, *Journal of the Science of Food and Agriculture*, 1973. Vol. 34 (2), pp. 249–52.

62 Marley, G., *Climate and the British Scene*, 1952, London, Collins.

63 Marot, G., Hybrid vines and the new viticulture, in Hyams (ed.), *Vineyards in England*. pp. 67–73.

64 Martin, H., *The Scientific Principles of Crop Protection*, 5th edn., 1964, London, Arnold. p. 110.

65 Martin, H., *The Insecticide and Fungicide Handbook*, 4th edn., 1972, Oxford, Blackwell.

66 Massel, A., *Applied Wine Chemistry and Technology*, 1969, London, Heidelburg Publishers Ltd. pp. 65–8, 155.

67 Miller, P., *The Gardeners Dictionary* . . . , 4th edn., 1743, London, Rivington. Gathering 8x (pages not numbered).

68 Moser, Lenz, *High Culture System*, n.d. (?1974), Krems, Donau.

69 Moser, Lenz, *Horticultural Abstracts*, 1975, Slough, Commonwealth Bureau of Horticulture and Plantation Crops. Vol. 45, No. 4786.

70 Muir, A., Vineyard soils in England, in Hyams (ed.), *Vineyards in England*. pp. 49, 102–8.

71 *Museum rusticum* (Anon. 'R-m'), 1776, London, XXXIX.

72 Musgrave, William, *Antiquitates Britano-Belgicae*, 1719, London. Vol. I, p. 192; vol. IV, pp. 11, 13.

73 *The Nautical Almanac*, 1968, Glasgow, Brown, Son and Ferguson.

74 Neal, D. A., Excavations at the Palace and Priory at King's Langley, *Hertfordshire Archeology*, 1973. pp. 31–5.

75 Nègre, E., and Françot, P., *Manuel pratique de vinification et de conservation des vins*, 1946, Paris, Flammarion. p. 203.

76 Ordish, G., A hundred years of lime-sulphur, *Agriculture*, 1951, London, Ministry of Agriculture. pp. 111–5.

77 Ordish, G., Wine, home-made but real, *Sport and Country*, 24 December, 1952. pp. 624–5.

78 Ordish, G., *Wine Growing in England*, 1953, London, Hart-Davis. pp. 101–5.

79 Ordish, G., The vine in spring, *The Field*, 10 March, 1954.

80 Ordish, G., *The Great Wine Blight*, 1972, London, Dent.

81 *Oxford Dictionary of the Christian Church*, 1974, Communion under both kinds, London, Oxford U.P.

82 Page, William (ed.), *The Victoria County History of Hertfordshire*, 1923. Vol. I, p. 343; vol. III, pp. 348, 352, 383, 424.

83 Pasteur, L., *Études sur le vin*, 1866, Paris.

84 Pearkes, G., *Growing Grapes in Britain*, 2nd edn., 1973, Andover, Hants, *Amateur Winemaker*.

85 Pegge, Samuel, Of the introduction, progress, state and condition of the vine in Britain, *Archeologia: or Miscellaneous Tracts*

Relating to Antiquity, 1804, London, Society of Antiquaries. Vol. I, LV, pp. 344–57.

86 Philipot, Thomas, *Villare Cantianum or Kent Surveyed*, 1659, London. pp. 93, 112.

87 Plott, R., *The Natural History of Staffordshire*, 1686, Oxford.

88 Pope, A., Essay on criticism, *The Poetical Works of Alexander Pope*, 1785, Glasgow, A. Foulis. Vol. I, p. 122.

89 Poulter, W. B. N., Some Aspects of Modern Viticulture in the English Climate, 1971, thesis (unpublished) for Diploma, Oenological Research Institute, Ockley, Surrey. pp. 54, 352.

90 Radnlov, L., Babrikor, D., *et al.*, High stem and wide inter-row cropping, *Horticultural Abstracts*, 1973, Slough, Commonwealth Bureau of Horticulture and Plantation Crops. Vol. 43, No. 5844.

91 *Review of Plant Pathology*, 1975, Commonwealth Mycological Institute, Kew, Surrey. Vol. 54, No. 2380.

92 Richmond, I. A., *Roman Britain*, 1956, London, Penguin.

93 Robertson, J., *Transactions of the Horticultural Society*, 1824, London. Vol. 5, p. 175.

94 Rohde, E. S., *The Story of the Garden*, 1932, London, Medici Society, p. 36.

94A Roques, J. F., Weed control in French vineyards, *Outlook on Agriculture*, 1976, Bracknell, Berkshire, ICI. Vol. 9, (I), pp. 30–4.

95 Rose, J., *The English Vineyard Vindicated*, 1666, London.

96 Seltman, C., *Wine in the Ancient World*, 1957, London, Routledge. pp. 91, 159–72.

97 Shakespeare, William, *The Tempest*. IV, i.

98 Shakespeare, William, *Hamlet*. II, i, 300.

99 Singer, C., *History of Technology*, 1956, Oxford, Clarendon Press. Vol. II, p. 137.

100 Slicher van Bath, B. H., translated by O. Ordish, *The Agrarian History of Western Europe*, 1963, London, Arnold. p. 132.

101 Smith, William, *Dictionary of Greek and Roman Antiquities*, 1859, London, Walton and Maberly. p. 945.

102 Somner, William, *The Antiquities of Canterbury*, *c.* 1650. pp. 81, 145, 170.

103 *Spectacle de la Nature*, Anon., trans. S. Humphreys, 8th edn., 1757, London.

104 Speechley, William, *A Treatise on the Culture of the Vine*, 1791, Dublin. pp. 234–72.

105 Strutt, J., *Doend Angel, or Manner of the Inhabitants of England*, 1775, London.

106 Sweet, H., *The Oldest English Texts*, 1885, London, Early English Text Society. p. 444, line 22.

107 Thearle, R. J. P., Some problems concerning the control of bird-

damage in Britain, *Proceedings of the 8th British Insecticide and Fungicide Conference, 1975*, 1976, Nottingham, Boots. Vol. III.

108 Thorn, William, *Chronicle* (14th century), Canterbury. Col. 2036.

109 Thoyras, Rapin de, translated by N. Tindal, *The History of England*, 1722, London. Vol. I, p. 21.

110, 111 Twyne (Twine), J., *Bolingdunensis, Angli, de Rebus Albionicis, Britannicis atque Anglicis, commentariorem, libri duo*, 1590, London. pp. 115, 116.

112 Viala, P., *Les maladies de la vigne*, 1887, Paris, Delahaye; Montpellier, Coulet.

113 Vispré, F. X., *A dissertation on the growth of wine in England to serve as an introduction to a treatise on the Method of cultivating vineyards in a Country from which they are at present entirely eradicated; and making from them good substantial wine*, 1786, Bath.

114 Walker, C. D., The cost of establishing a vineyard in the United Kingdom, *English Vineyards Association Journal*, 1975. Vol. 9, pp. 44–5.

115 Ward, J. L., Propagation of protected varieties, *English Vineyards Association Journal*, 1974. Vol. 8, pp. 14, 15.

116 Ward, J. L., Scourge of the vineyards, *English Vineyards Association Journal*, 1974. Vol. 8, pp. 48–53.

117 White, K. D., *Roman Farming*, 1970, London, Thames and Hudson.

118 Wilkins, David, *Leges Anglo-Saxonicae*, 1721, London.

119 Wilson, D. M., *The Anglo-Saxons*, 1971, Harmondsworth, Penguin. p. 88.

120 Wilson, O. S., *The Larvae of the British Lepidoptera*, 1880, London, Reeve. pp. 35–7, 88, 244–5.

121 *World Weather Records, 1941–1950*, 1959, Washington, D.C., U.S. Dept. of Commerce.

122 Young, Jessica, Vintage wine a penny a bottle, *Farmer and Stockbreeder*, 30 November–1 December, 1954.

Index